Song Full of Tears

MAO QILING, *Chronicler of Xiang Lake*
(1623–1716)

Song Full of Tears

Nine Centuries of Chinese Life at Xiang Lake

R. Keith Schoppa

A Member of the Perseus Books Group

Published with assistance from the Ziegler Family Faculty Research Fund for the Humanities, Valparaiso University, Valparaiso, Indiana.

First published by Yale University Press, 1989.

Westview Press books are available at special discounts for bulk purchases in the United States by corporations, institutions, and other organizations. For more information, please contact the Special Markets Department at The Perseus Books Group, 11 Cambridge Center, Cambridge MA 02142, or call (617) 252–5298.

Published in 2002 in the United States of America by Westview Press, A member of the Perseus Books Group, 5500 Central Avenue, Boulder, Colorado 80301–2877, and in the United Kingdom by Westview Press, 12 Hid's Copse Road, Cumnor Hill, Oxford OX2 9JJ

Visit us on the World Wide Web at www.westviewpress.com

Designed by James J. Johnson and set in Sabon types by Graphicraft Typesetters Limited, Hong Kong.

A Cataloging-in-Publication Data record is available at the Library of Congress

ISBN 0–8133–4020–9

The paper used in this publication meets the requirements of the American National Standard for Permanence of Paper for Printed Library Materials Z39.48–1984.

10 9 8 7 6 5 4 3 2 1—05 04 03 02

FOR

Kara, Derek, and Heather

—to memories of our Hangzhou autumn

All things fall and are built again,
And those that build them again are gay.

Two Chinamen, behind them a third,
Are carved in lapis lazuli,
Over them flies a long-legged bird,
A symbol of longevity;
The third, doubtless a serving-man,
Carries a musical instrument.

Every discoloration of the stone,
Every accidental crack or dent,
Seems a water-course or an avalanche,
Or lofty slope where it still snows
Though doubtless plum or cherry-branch
Sweetens the little half-way house
Those Chinamen climb towards, and I
Delight to imagine them seated there;
There, on the mountain and the sky,
On all the tragic scene they stare.
One asks for mournful melodies;
Accomplished fingers begin to play.
Their eyes mid many wrinkles, their eyes,
Their ancient, glittering eyes, are gay.

—WILLIAM BUTLER YEATS,
"Lapis Lazuli"

Contents

Illustrations

FIGURES

MAPS

Preface

Xiang Lake: a six-thousand-acre reservoir ten miles from scenic West Lake in the city of Hangzhou. Nothing of national importance either in empire, republic, or people's republic ever occurred there; no disastrous rebellions or other social disturbances ever began there; no conferences of national significance were ever held there; no figure of national scope ever rose from its shores; none of China's preeminent poets wrote of its beauty. In the vast sweep of China's history it seems, on the whole, quite forgettable, a lake of little significance. Why then should we spend time at Xiang Lake?

The story of Xiang Lake is a drama of the human struggle for life and for control of the environment. It opens a nine-hundred-year window on Chinese society, especially on individuals—officials, elites, commoners—whose lives were involved in or played themselves out around Xiang Lake. In a nation where "faceless crowds" of millions, and now a billion, people have become a cliché, the lake provides the human face of men and women dealing with problems of daily life. In the geographical vastness that is China, the story of Xiang Lake grounds us in a particular limited space to which we can become accustomed and in which we can see change occur over time.

More than simply providing a dramatic story, the history of the lake offers a view of how Chinese society "worked" in times of environmental and military crises; how Chinese reacted to changes, threats, and opportunities; how they dealt with one another and why they did so in particular ways; and how they responded to the world of nature and the environment. Because of the length of the lake's history, we may also ponder questions about the continuity of the Chinese views of life, society, and nature through the centuries. In a world where intercultural understanding is often rudimentary at best, the depiction of how Chinese have dealt with their problems and self-interests gives us a greater sense of the social dynamics of Chinese culture. It seems clear, for example, that many contemporary observers of China incorrectly attribute various Chinese attitudes and approaches to the effects of totalitarianism when, in fact, those attitudes and approaches stem from longstanding sociocultural dynamics.[1]

The story of the lake contains a number of themes that make it conspicuously relevant for some concerns in the last years of the twentieth century. Struggles over the environment and ecology have existed for centuries in China in the age-old effort to balance the number and the needs of people with available land and water resources. Linked to but also independent of these concerns has been the enlarging sphere of public claims on private interests, including the expanding role of the state in matters once perceived as private. Such issues arose, as we will see, in the day-to-day struggle that marked the lives of Xiang Lake area residents.

The story of the lake has much to teach us about the Chinese and their society, past and present. It is an antidote to the politics-determined national regimen that petrifies our view of the past into emperor's reigns, dynasties, republic, or people's republic and reduces historical inquiry and explanation to exploring the past already bound by such rubrics as "feudalism," "revolution," and "modernization." The poison of these approaches is their destruction not only of the day-to-day particularity of Chinese life but also of basic social, cultural, and political continuities and discontinuities that become obscured by various periodizing and thematic assumptions.

For all its importance of bringing us to the particular, on another level Xiang Lake can be seen as a symbol of traditional Chinese civilization. Created at a time often referred to as the beginnings of "early modern" history, the lake came to be seen as the locality's heritage, the legacy of the founders and preservers of the lake.[2] While other lakes in the area were reclaimed, recreated, and reclaimed again, Xiang Lake, though altered, remained functioning over eight centuries. Like the centrality of Confucian ethical values in the sociocultural realm, the lake came to stand for the economic center of the life of the area's people. Heroic protectors of this legacy dealt with varied, though repeated, threats of social-ethical heterodoxy: official benevolence turned malevolence; cohesive family interests turned outward to arrogant greed; public-spiritedness turned to bald self-interest. Like many Chinese traditions in the nineteenth century, the lake was seriously damaged by natural disaster and outside invaders. Like the traditional political and ethical orthodoxy, the lake was largely destroyed in the twentieth century.

Traditional Chinese culture was built on remembrance of the past: the past showed the way to the future. Macropolitical remembrances took the form of epochal or dynastic histories; local remembrances were histories of provinces, prefectures, counties, or geographical units (like Xiang Lake), as well as nonverbal memorials such as shrines to outstanding local leaders; family remembrances included genealogies and ancestral rituals and shrines; individual remembrances consisted of biographies, personal chronologies, and essays. With a plethora of written records, shrines, memorial arches, ancestral tablets, and an acute awareness of ancestors, the Chinese present was surrounded by and infused with the past. The environs of Xiang Lake had numerous sites with historical relics or connections to figures of some historical significance.

We too are now rememberers of the past of Xiang Lake and the people around it. Our remembrances are shaped by the images and approaches of earlier historian-rememberers: Gu Zhong (twelfth century); Zhang Mou (fourteenth); Mao Qiling (seventeenth); Yu Shida (eighteenth); and Zhou Yizao (twentieth). Our understanding of the lake's past is only as accurate as the accounts of these earlier historians. The narrative with documents in the accounts of Gu, Zhang, and Mao set forth a particular point of view: preserving the lake was necessary; reclaimers were ipso facto self-interested and thereby unethical. That becomes our image as well, though not uncritically: it is important to note that other historical sources—essays and poems, in particular—corroborate this historical interpretation. The approaches of Yu and Zhou are notable in contrast. The "praise-blame" thrust of much of traditional Chinese historical writing is largely absent in Yu's factual comparison of the lake in the 1790s with its original state and in Zhou's twentieth-century compilation of documents setting forth the arguments of reclaimers and defenders of the lake. In these cases, other briefer accounts in essays and newspapers provide additional information that helps inform our point of view.

Sources for this story of the lake are the historical accounts of the men mentioned above, county histories, documents on the lake, essays by county figures, genealogies of key county lineages, and newspaper and government reports. Poetry, the highest literary art of the traditional Chinese literati, also provides a valuable source.

Each chapter is preceded by several pages of a "view," brief sections with several objectives. First, each view imitates in prose the traditional literati practice of selecting special sites of scenic beauty (often located in a garden) and recording them in painting or poetry. Scholar-gentlemen in Xiaoshan county had, likely with some pretension, denoted eight such spots, thereby equaling but not surpassing the well-known eight sites of Beijing.[3] A second objective is to place the reader of the lake's history on key sites in Xiaoshan county and the lake area: each view describes what one could have expected to see at a particular time from that vantage point. Finally, each "view" adumbrates the themes and occasionally the episodes that are significant in the chapter which follows; they introduce our vision of Xiang Lake at one time in its history.

In the past centuries, countless important visitors have come to the lake and its environs. Some came on business, inspecting the sites and learning about lake functions; many came for pleasure and relaxation in the natural world of lake, springs, and mountains; some passed through en route to other destinations; and a few came ceremoniously in their positions of power. Most lake visitors were probably not as enthusiastic as Wu Shu, a late-fifteenth-century county official who described himself as deeply moved, even ecstatic, over the vista of the lake amid the mountains.[4] Famous philosopher Wang Yangming marked his visit in the sixteenth century with a poem.[5] Chairman Mao Zedong visited the lake region in the late summer of 1959, a visit remembered for its impact on the local population rather than for any notable effect on Mao.[6]

We too are now visitors at Xiang Lake. Of central importance in the lives of thousands of Chinese, it becomes both the central character and setting of our story.

Acknowledgments

My fascination with Xiang Lake and my desire to tell its story began several years ago when I came upon the stunning tale of vengeance related in chapter 2. Pursuing Xiang Lake's story to the present repeatedly confirmed to me that it had much to disclose about the development and dynamics of Chinese society and culture.

During the years of research and writing on this project, I have incurred various debts to numerous people and institutions. Clearly my story of the lake would not have reached its proper conclusion had it not been for the great generosity of Professor Chen Qiaoyi of the Department of Geography at Hangzhou University. Professor Chen gave freely of his time and energy during my stay in Zhejiang in autumn 1986, providing me with sources, arranging interviews and library privileges, and arranging and accompanying me on an important research trip to Xiang Lake sites in Xiaoshan county. His personal knowledge of the lake area provided me an immense wealth of additional data and saved me from numerous errors. Those errors which remain are, it should be noted, my responsibility. In addition to Professor Chen, I want to thank Professor Lu Yichun and Mr. Que Weimin of the Hangzhou University Department of Geography for their conscientious help and effort in my behalf. For general assistance at Hangzhou University, I also want to thank Chen Xinqi, Deputy Director of the Foreign Affairs Office, and Vice President Xia Yuejiong. In Xiaoshan county, special thanks should go to Lou Gengyang, vice editor of the 1985 draft of the county gazetteer; Dong Mingzhi, chief engineer of the county water control office; and Yang Jun, teacher at Xiang Lake Normal School.

I owe another major debt of gratitude to the East-West Center in Honolulu. I was fortunate enough to begin my study of China there as a graduate student from 1966 to 1968 and to return from January to July 1986 as an Alumnus-in-Residence Fellow, during which time I completed a draft of the first seven chapters. I especially want to thank Gordon Ring, Sumi Makey, Glenn Shive, and June Sato in the Office of Student Affairs and Open Grants for their help and hospitality. For comments and suggestions on various aspects of the project, I offer my gratitude to other East-West Center fellows: Hwang Kwang-kuo, Lucien Miller, and William Whiteside; and to University of Hawaii professors Harry Lamley, Daniel Kwok, and Brian McKnight. For various advice and help in some translation, I would like to thank Fan Lei.

For reading the manuscript and providing cogent criticism and friendly support, I am indebted to Ernest Young, William Rowe, Mark Schwehn, and Richard Vuylsteke. In addition, the following colleagues provided comments, assistance, and support at various stages of the project: Parkes Coble, James Dew, Bradley Geisert, John Israel, Yoshinobu Shiba, Rubie Watson, James Watson, and Philip West. I wish also to thank the typists for the manuscript: Berta Acojido, Candace Conrad, Deanna Ho, Christine Ideue, Colleen Maielua, and Beth Schoppa. Yale University Press prepared the chapter maps of Xiang Lake, and Ken Keifenheim of the Valparaiso University Department of Geography drew the others. At Yale University Press, I want to express my gratitude to Chuck Grench and Laura Jones Dooley.

I am grateful for the financial assistance that facilitated the research and writing of the book. Funds were provided by two grants from the National Endowment for the Humanities and the East-West Center Alumnus-in-Residence Fellowship. For time and support to complete the project, I owe deep gratitude to Valparaiso University for awarding me the University's Research Professorship, 1986–1987, and for providing funds for cartographical and publication expenses, the latter from the Ziegler Family Faculty Research Fund for the Humanities.

Finally, I offer my deep thanks and love to my wife, Beth, whose name appears above as typist but who always provided much more in countless ways to the project and especially to our life beyond Xiang Lake. To Kara, Derek, and Heather, to whom this book is dedicated, I am grateful for compelling me to see my research in the broader, if at times more mundane, perspective of family life and priorities. In turn, I hope that their introduction to the world of China through my research and their experiences abroad can bring another meaningful perspective to their lives.

Periods in Early Modern and Modern Chinese History

Northern Song	960–1126
Jurchen/Mongol rule in north Southern Song in south	1127–1279
Yuan (Mongols)	1279–1368
Ming	1368–1644
Qing (Manchus)	1644–1912
Republic	1912–1949
People's Republic	1949–

Major Personages in the Life of Xiang Lake, Xiaoshan County

CAI PANLONG — Scholar-poet of the lake, late thirteenth century

CAI ZHONGGUANG — Scholar, poet, and hermit of the seventeenth century

GU ZHONG — Late twelfth-century county magistrate who opposed numerous encroachers on the lake

GUO YUANMING — Early thirteenth-century county magistrate who delineated boundary between lake and land

HAN SHAOXIANG — Early twentieth-century dike manager who opposed lake reclamation

HE JING — He the Filial Son, late fifteenth–early sixteenth century

HE SHUNBIN — Late fifteenth-century censor who tried to stop lake encroachment by powerful lineages

HUANG YUANSHOU — Late nineteenth–early twentieth-century land developer

INTENDANT ZHANG — Twelfth-century lake villain

LAI JIZHI — Scholar-official, poet, and local reformer in the seventeenth century

LAI QIJUN — Late eighteenth-century protector of the lake

MAO QILING — Lake chronicler and activist, seventeenth and eighteenth centuries

SUN DEHE — Companion of Cai Panlong on outing in 1271

SUN KAICHEN — Lineage leader who attempted to build a cross-lake bridge, 1689

SUN QUAN	Lineage leader whose actions prompted the Censor He affair
SUN XUESI	Official and lineage leader who built the Cross Lake Bridge, mid-sixteenth century
SUN ZHAOWU	Lineage leader involved in lake encroachment, early sixteenth century
WANG XU	Late eighteenth–early nineteenth-century lake leader and ally of Yu Shida: involved in dike and sluice-gate repair
WEI JI	Mid-fifteenth-century lake poet and reformer, often linked in importance with lake creator Yang Shi
XIANG YU	Aristocratic contender for the throne from the Shaoxing area, second century B.C.
YANG SHI	Early twelfth-century county magistrate who prompted the creation of Xiang Lake
YU SHIDA	Late eighteenth-century scholar-investigator of the lake's condition and restorer of its facilities
ZHANG MOU	Fourteenth-century lake restorer; builder of Shrine for the Four Senior Officials
ZHAO SHANQI	Mid-twelfth-century county assistant magistrate who dealt with a variety of lake problems
ZHOU REN	Twelfth-century imperial military commissioner involved in reclaiming area lakes
ZOU LU	Late fifteenth-century county magistrate and conspirator in the Censor He affair

Note on Pronunciation

In transliterating Chinese, I use the *pinyin* system of romanization. Pinyin names are generally pronounced as written with the following exceptions:

c: At the beginning of a syllable, *c* has the sounds of *ts*; thus Cai Zhongguang is pronounced Tsai Zhongguang.

q: Q sounds like a hard *ch*; thus Mao Qiling is pronounced Mao Chiling and Sun Quan is pronounced Sun Chuan.

x: *X* is pronounced halfway between *s* and *sh*, in another romanization written as *hs*. Sun Xuesi is thus pronounced Sun Hsuesi.

The only exceptions to *pinyin* usage are Sun Yat-sen and Chiang Kai-shek, the forms of which are best known in Western writings and which give Cantonese, not Mandarin, pronunciations of the Chinese characters.

I

VIEW

From North Trunk Mountain, Late Twelfth Century

CHAPTER

The Beginnings, 1112–1214: Four Servants of the People

VIEW: *From North Trunk Mountain, Late Twelfth Century*

The panorama from North Trunk Mountain's Jade Peak was dominated by water. In all directions, in the early summer months, the very land itself seemed afloat. The rice seedlings showing above the water level of the paddies gave the landscape a delicate emerald sheen. Lakes and streams fed a weblike network of canals and drainage ditches with sustaining water flowing to every bounded plot.

To the west and beyond the lake to the south was the broad river the Qiantang, at its widest almost two miles across, at its narrowest, at least half a mile. Flowing from headwaters in the mountains of southwestern Zhejiang, it deposited its heavy load of sediment from constantly shifting currents as it made three sharp bends around the peninsula that was the western part of the county. Reports of the rapid changes brought by its currents and alluvium continually startled area residents and travelers alike. In the early days, one of the eight scenic wonders of Xiaoshan county had been the sound of the river tide against the Raksha rock, a large boulder with precipitous slopes, which, like the Buddhist demons for which it was named, devoured men as it wrecked their boats. In the tenth century the river, as if showing mercy, began to cover it with silt; by the sixteenth century the tremendous rock could no longer be seen.[1] The constantly changing river prevented Hangzhou, the Zhejiang provincial capital—in the Southern Song dynasty (1127–1279), the imperial capital—from ever developing as a seaport.

At times the existence of the river and its extreme unpredictability provided the county with protection from the west against marauding enemies. In several cases of such outer threat, local wisdom credited the spirit of North Trunk Mountain with the county's escape. The mountain, it was said, held the spirit of Li Di, a general who had forsaken the oppressive Qin dynasty (221–207 B.C.) to follow into battle the aristocratic warrior Xiang Yu, a native of Shaoxing, east of Xiaoshan. Killed in battle, Li was buried on the mountain; he was said to have embodied a special virtue that allowed him both to see the misrule of the Qin and to serve Xiang loyally. For this reason, local people had constructed a temple for him on the mountain, where every year on the fourth day of the first lunar month sacrifices were offered.[2]

The case of the fearsome bandit-rebels under the leadership of Fang La, a disaffected owner of a lacquer tree plantation southwest of Xiaoshan, was often described to prove the spirit's efficacy. Poised in 1121 on the western bank of the river only about six miles from the county seat, the rebels were a particularly bloodthirsty lot.[3] Not only were they animated by a bitter hatred of all officials, depicted as exactors of a never-ending series of taxes, but their grievances were undergirded by Manichean religious ideas that treated killing as a positive vehicle for salvation.[4] In addition, the rebel army adopted psychological tactics that unnerved their opponents. Donning clothes of bright colors, painting their faces, and fashioning tall likenesses of wild animals and phantasmic creatures from skins, costumes, and mechanical devices, they strode into battle with loud cries, panicking the local inhabitants. Fearful that the river might not keep this scourge away, the populace offered prayers to the spirit of North Trunk Mountain. That night a strong wind blew miraculously from the northeast, wrecking rebel boats and preventing the river crossing. Apparitions of a massed army with a giant among them dressed in tiger skins were said to have appeared on Xiaoshan's riverbank. The spirit of the dead general had mimicked the rebels' own techniques. The astonished rebels dared not cross the river and attack the county.

For Xiaoshan inhabitants during the Song dynasty, the natural world was not simply the scenic beauty of the tall pines on North Trunk Mountain, renowned as one of the county's scenic spots. Uncontrollable natural forces also provided a continual challenge. That Li's ghost chose to dress in tiger skins reflected the fact that as recently as the century before man-eating tigers had their bloody lair less than half a mile from North Trunk Mountain.[5] Wild animals, of course, could not compare with the awesome power of nature at its most terrifying in the seemingly unstoppable power of raging floods. When heavy mountain storms to the south made the Qiantang River a raging torrent, it slammed with devastating force into the Xiaoshan bank between Fish Lake and Wen Family Dike. Chinese legend tells of the great King Yu, who through his flood control efforts saved the Chinese from becoming fish. Yu's home has traditionally been associated with the present area of Shaoxing prefecture in which Xiaoshan county is located,[6] and the danger and reality of frequent flooding here made that understandable. The view from North Trunk Mountain revealed that the terrain only exacerbated the flooding problem. Land sloped to the sea from the mountains to the south. Instead of ultimately draining into the sea, however, much of the water remained in low-lying areas that were rapidly transformed to swamp. It was Yu's contribution to set up an intricate network of drainage canals to deal with this problem.

Between North Trunk and the bay to the north lay alluvial flats deposited by the river and peopled by squatters in the eternal search for more productive, if treacherously impermanent, land. From the sea came the salt fields with a commodity so important in Chinese society that its production and distribution early on became a government monopoly. But like some jealous water god ruling by whim, the sea could turn its fury on those alluvial flats and salt fields, taking what it had given. Storms frequently sent seawater towering in destructive waves over the land, leaving its deadly saline residue.

Each year at the time of the Mid-Autumn Festival, the top of North Trunk Mountain itself was the destination of crowds of sightseers who climbed to watch one of the sea's most spectacular water displays: the arrival of the tidal bore pushing up the river's estuary. The tide rushed in ten to twenty feet high at some thirty miles per hour; its roar could be heard thirty to forty minutes before it reached Hangzhou. Such displays of the force of water brought early Chinese to respect the god of the tide, and well it might: in September 1132, several hundred spectators on specially constructed wooden stands on the river's bank were swept to their death in the surf when the tide unexpectedly destroyed their viewing place.[7] The Ningqi Temple at Xixing was established to sacrifice to the spirit of the tidal bore.[8] Imputing natural phenomena to the designs of spirits, like the story of Li Di's ghostly machinations to forestall Fang La's attack, was a Chinese effort to explain awesome forces in more understandable, if fabulous, human terms.

Flowing west directly past the base of the mountain was the Grand Canal, carrying from Shaoxing, about twenty-five miles to the southeast, many boats loaded with goods from south China. After reaching Xixing on the Qiantang River's Xiaoshan bank, they would be ferried to Hangzhou.[9] Dubbed "the finest and noblest in the world" when Marco Polo visited it in 1286, this city was clearly visible from the pavilion on North Trunk's Jade Peak.[10] Outside the city wall on the west edge of the city was West Lake, which provided drinking water for the city's inhabitants but was known chiefly for its natural beauty and for the poetry, painting, and convivial social gatherings it stimulated.

In the water-dominated world of Xiaoshan perhaps most notable from the mountain were the lakes. The larger Ning-Shao (Ningbo-Shaoxing) Plain in which Xiaoshan is situated is bounded by ocean (the East China Sea on the shore of Ningbo prefecture), river (the Qiantang alongside Shaoxing prefecture), Hangzhou Bay, and mountains to the south (see map 1). In the Song dynasty, it had 217 lakes—a large lake every twenty square kilometers—the greatest density of lakes in all China.[11] Lakes served as retaining basins to contain inordinate rainfall and mitigate the flooding problem. More important, they were reservoirs, collecting precious water for irrigating paddies when little rain fell. For this is the irony of life in Xiaoshan: in a world dominated and constrained by the abundance of water, the actual problem year in, year out tended to be drought. When a local official was asked in 1353 to climb North Trunk Mountain to the general's temple, it was to pray for deliverance not from bandit-rebels but from drought.[12] The plight of the unfortunate Chinese farmer was literally flood or famine. Lakes became a way to help balance natural precipitation.

Many lakes in the area had been constructed for irrigation. Mirror Lake, for example, a large reservoir in Shanyin county to the east of Xiaoshan, had been created in A.D. 140 by building embankments around a number of natural springs.[13] Smaller lakes in Xiaoshan county—Shooting Star Lake, Catalpa Lake, Dripping Melon Lake—had been built in similar manner for such a purpose. Xiang Lake, the largest in the county, was built to contain the water from many springs in the surrounding mountains and to retain rainfall as well.[14] From the view atop North Trunk Mountain, it stretched to the south, roughly in the shape of a gourd, narrow in the north, rounded and fuller to the south. Less than a third of a mile farther south flowed the great river, menacing in its potential for destructive floods.

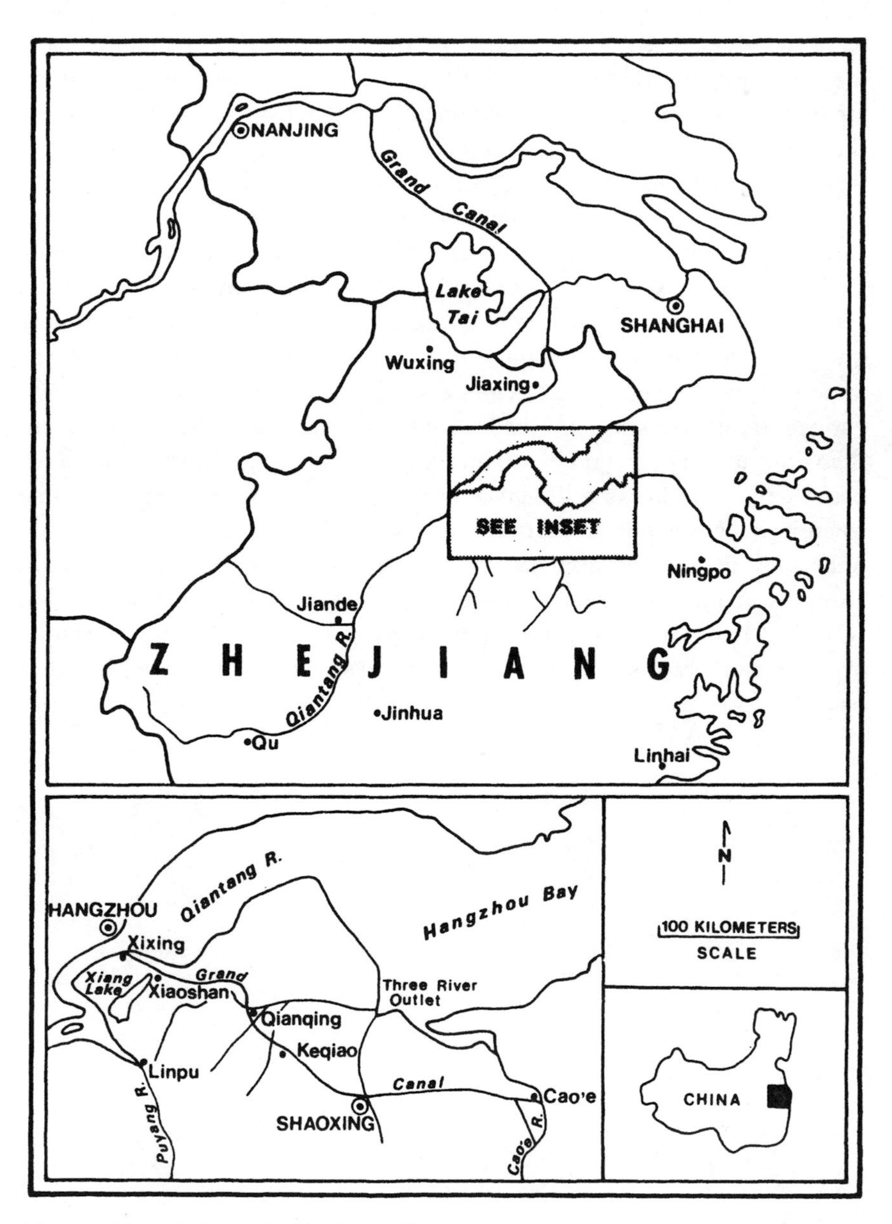

Map 1. Xiang Lake in Its Regional Context

The concept of *fengshui* (literally, "wind-water"), or geomancy, shaped the Chinese response to the natural world. Buildings and gardens were constructed and gravesites placed only in favorable geomantic spots ideally where natural topographical conditions combined an appropriate amalgam of *yin* and *yang*, dichotomous but complementary forces. Yang, the male force in the universe, was symbolized in the natural world by mountains, the heavens (where the sun made its daily course), and the south (the source of the life-giving warmth of the sun).[15] Yin, the female force, had among its many manifestations, valleys, water, and the north.

The historical descriptions of North Trunk Mountain associate it, in its protective qualities, with the county itself; its base was located only a few hundred paces from the county seat, where the representative of the imperial government acted as chief county official, or magistrate. To the yang or life-giving south of North Trunk lay Xiang Lake. More significant than the mountain, symbol of the county and alleged protector of its people, the lake became life-giver and sustainer to its surrounding nine townships. It is reported that the famous Song dynasty poet Su Dongpo once said that Hangzhou had West Lake like a person has eyes. In contrast, in 1501, Liu Zhang, a Ming dynasty scholar-official, wrote that Xiaoshan had Xiang Lake like a person has a stomach. "If one covers the eyes, he can't see; but if one's stomach has serious problems, he can't live."[16]

CHAPTER: *The Beginnings, 1112–1214: Four Servants of the People*

In the early winter of 1271, two young men, Cai Panlong and Sun Dehe set out from the town of Xiaoshan to a pavilion about five miles south on the shore of Xiang Lake and in clear view of the Qiantang River.[1] Both were involved in studies in Xiaoshan and wanted a break from the routine; in the mode of Chinese gentlemen, they had decided to spend their leisure viewing the scenery from the pavilion, playing music, and drinking. Sun had invited Cai to the pavilion; he apparently felt a special tie to the place because his older brother, Sun Delin, widely known for his literary skills, had been commissioned to pen a new name for the pavilion shortly before he suddenly and unexpectedly died. The pavilion had been built by a member of the Han family, whose lands were to the southeast of the lake and who had been accustomed to referring to it as the Officials' Pavilion, since it was generally used for the convenience of officials coming from and going to Hangzhou. Evidently someone in the Han family decided a new name written on an identifying tablet by an illustrious litterateur would dignify their pavilion. The new name: Pavilion for Viewing the River and the Lake.

It was cold. The day alternated between periods of sunlight and mist, and as the two men leaned on the pavilion's railing, occasional gusts of wind whistled through the structure. We do not know if they played the four-stringed guitar, how much they drank, or how long they stayed. What survives from this outing, reflecting the Chinese love of natural beauty, is

Cai's reverie on the lake scenery and the view from the pavilion. In his prose, the lake, renowned as one of the eight scenic beauties in the county, displayed autumnal and winter images that, though this outing bespoke no outward sign of mourning, must have reminded Cai of the brevity of existence in light of the recent death of his friend's brother. The images of nature, especially those of the lake, seemed to overwhelm Cai.

> In the sparse forest, leaves laden with frost mask the sun. The heavens are studded with rosy clouds. The reeds along the shore and the fishing boats undulate in the waves blown by the west wind; on the sandy beach, the shadows of wild geese; swirling waters around the shoal—images of untamed wilderness.
>
> All around, cloudy mists accumulate in the skies while the slanting rays of the sun shimmer like gold on the lake. The sound of waves from the river assaults my ears—I seem to be hearing several kinds of music and singing. I am overwhelmed. The appearance of the water invades my senses. I have lost all track of time. Images, pictures come and go—one forgotten as it is replaced by the next. I am both amazed and confused.
>
> The waves play while wild ducks, tier upon tier, dance. . . . The distant mountains are bright in the mists; a waterfall cascades down the cliff. The scenery is rare and wonderful; on the shore a twisted pine together with a tiger-shaped stone; tall bamboos surround an old plum tree. In the distance a sail quickly falls while fishhawks skim over the water.
>
> Xiang Lake's unrestrained, heroic spirit pervades all; and all this is pressing in on my eyes here in this pavilion, fixing my gaze.
>
> Imagine! A year has four seasons—four different views. If it is like this in winter, imagine what it is like in spring, summer, and fall. Whether it will be clear or cloudy is not certain, but imagine if it is like this on a clear day what it is like when the wind blows strongly and there are clouds and rain. If it is like this in the day, imagine what it will be like at night with a bright moon glittering on the waves.

Cai's celebration of the vitality and beauty of the lake as seen from the pavilion pleased Sun, who grasped his hand enthusiastically and later remarked that Cai had given the name of the pavilion a proper descriptive record. If the lake was indeed to the county as a stomach to a body, it also obviously displayed some of the quality of eyes.

Re-creating the Lake: The Contribution of Yang Shi

Xiang Lake was constructed as a reservoir in the closing decades of the Northern Song dynasty (960–1126) in a site where there had once been a natural lake.[2] We know little about the earlier natural lake. Called the Lake West of the Wall (the wall of the county seat), it flourished from approximately the third to the ninth centuries as an appendage of an even larger lake to the west, Fish Lake (see map 2A). But the Puyang River flowing past it silted up its outlet to the larger lake, and little by little the Lake West of the Wall was reclaimed and became dry.[3] Farmed for two to three centuries, the former lakebed was natural lowland and subject to frequent floods. In part for this reason, proposals to turn it back to lake were heard by the late eleventh century.

There were other more compelling reasons to consider returning it to lake. Beyond the lakebed itself, there was great concern about the persistent problem of drought in nine surrounding townships. And in two low-lying townships north of the county seat, the solicitude was the periodic floods that poured from higher altitudes. The proposed reservoir would serve the dual function of flood-control facility and irrigation source. The Song court gave local elders permission to discuss construction of such a lake.[4]

But discussions brought no local consensus. Wealth largely determined elite status during the Song, and wealth was at stake in the proposal to build a lake covering almost nine square miles.[5] Some of the rich would have to give up their land for the lake at the very time when the rich in other nearby areas were reclaiming lake land as an appealing investment opportunity.[6] The idea was dropped.

But in the period 1107–1110, unnamed people in the county again asked for the re-creation of a lake. In 1112 an energetic county magistrate, Yang Shi, called a meeting of village elders (*qilao*) of the nine townships, men whose roles usually included caring for public works and supporting agriculture.[7] The detailed records of this important meeting have been lost. We do not know who supported or opposed the idea. There is also unfortunately no indication of how the landowners of the proposed lake land were induced to turn over private property for what might be called the gain of the Xiang Lake region. Patriarchal lineages like the Zhan lineage, five miles west of the county seat, and the Zhou lineage, eight miles to the south, had in the past given up productive land to form reservoirs for their lineages.[8] But Xiang Lake was built on a far different scale and for a far more heterogeneous community. With a circumference of about twenty-three miles, the lake covered over 37,000 mu, or about 5,600 acres (almost five times larger than West Lake), and was intended to irrigate almost 147,000 mu, or about 22,250 acres, of paddy land.[9] Mao

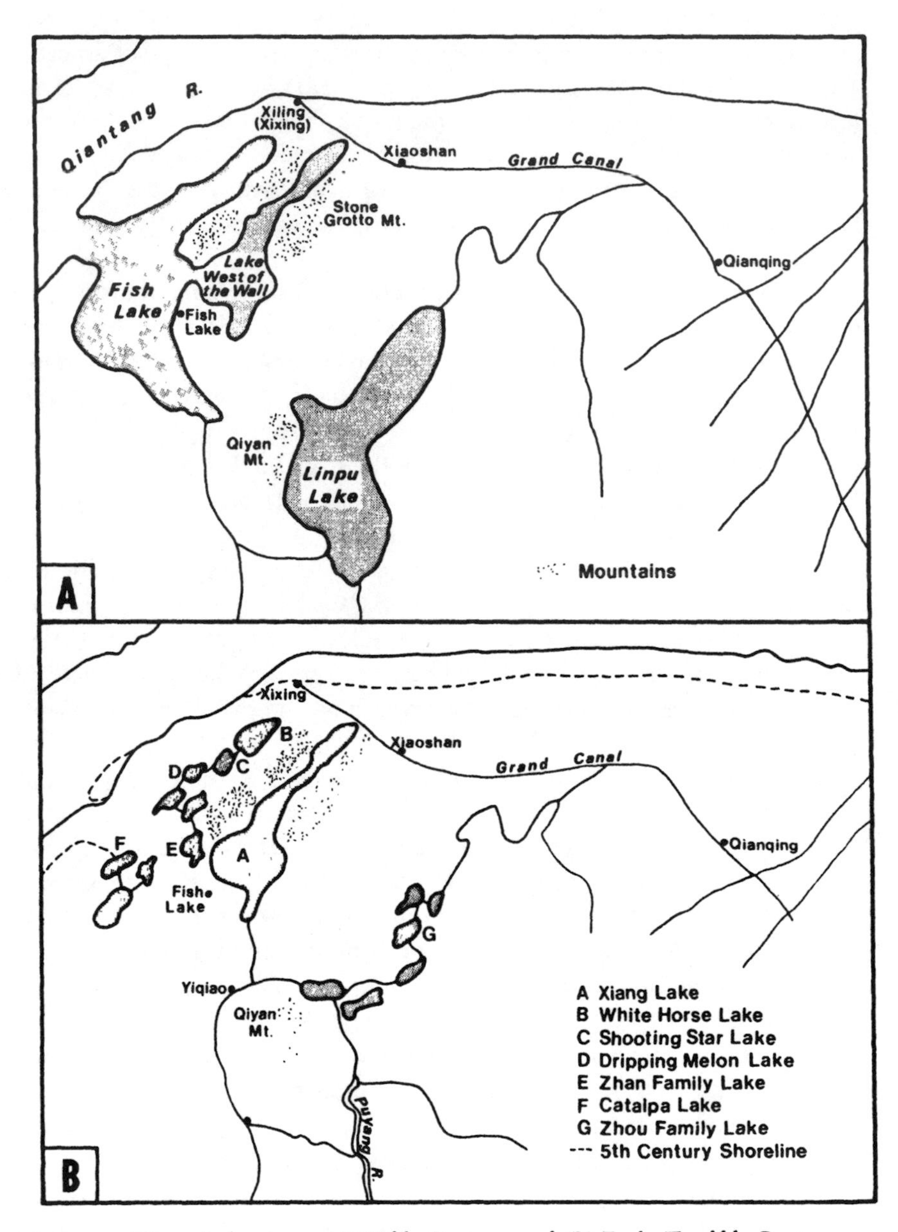

Map 2. Xiang Lake Area, (A) Fifth Century and (B) Early Twelfth Century

Qiling, the early Qing chronicler of the lake, provided no clues to the general situation or the motives of the landowners—whether they were forced to accede to the formation of the lake because of political weakness or were willing to assent because of economic desperation. Mao dealt only with the details of land tax policy. The state's interest in the lake's establishment centered mainly on taxes, which it did not want to lose from the productive cropland now covered by lake water (16,000 mu out of the 37,000 total). The elders of the nine townships satisfied the government's concern by agreeing to add an amount per mu to their land tax to cover what would have been collected on the cropland.[10]

The reservoir was built with mountains to the east and west serving as natural dikes along the length of the lake. Constructed dikes were necessary only in the north and south. The dikes were obviously higher than the lake; the lake was about 1.5 meters higher than the surrounding land. Water simply flowed from outlets in the dikes. The one-and-a-half mile southern dike had four irrigation outlets into two townships—Xinyi and Xuxian—which contained more than one third of the total acreage irrigated from the lake (see maps 3 and 4). The southeastern dike, two-thirds of a mile long, included three outlets draining into the four eastern townships. Through these seven outlets came water for over 80 percent of the total acreage irrigated from the lake. The major dike in the north, running about two-thirds of a mile from near the county seat to Chrysanthemum Mountain, had five outlets that provided water to three townships in the north. In addition to these dike outlets, six were located at various places along the mountains to the west and southwest of the lake.[11] A dike chief (*tangzhang*) was assigned to each outlet to oversee and keep it in good operating condition. His job was crucial in the equitable allotment of water. Also with authority in the regulation of the lake were "upper households" (*shanghu*), the wealthiest in the area, who, under the Song dynasty socioeconomic grading system, bore heavy local service obligations.[12]

For his role in re-creating the lake and thereby benefiting those around it, the magistrate Yang Shi won much acclaim.[13] Using methods not detailed in the sources, he initiated the project, dealt with the lake's landowners, oversaw dike construction (with its necessary gathering of laborers and supplies), and set down allotment regulations for the eighteen irrigation outlets. For these accomplishments the nine townships remembered Yang as one of the four "senior officials" of the lake. The remainder of this quartet—Zhao Shanqi, Gu Zhong, and Guo

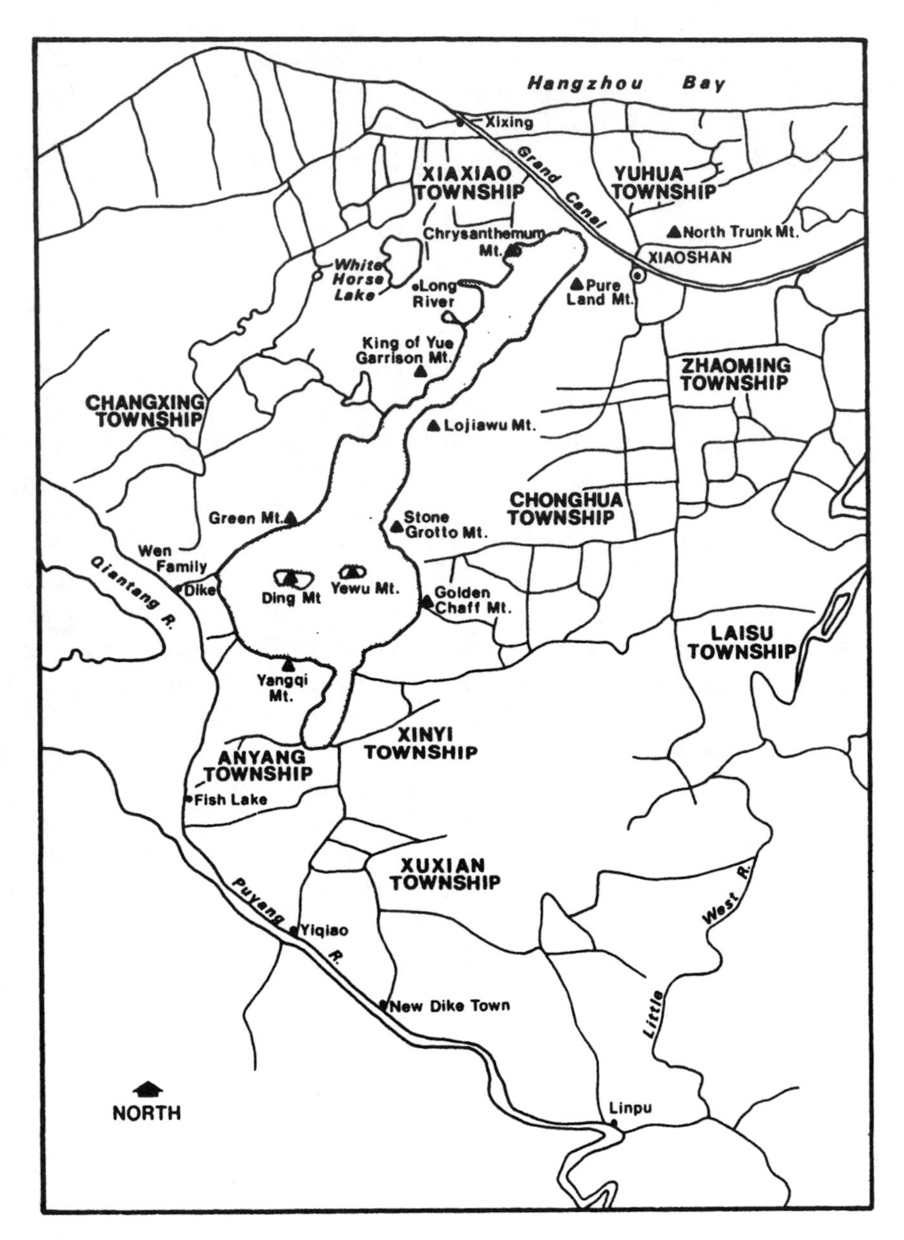

Map 3. Physiographic and Political Features of the Xiang Lake Area

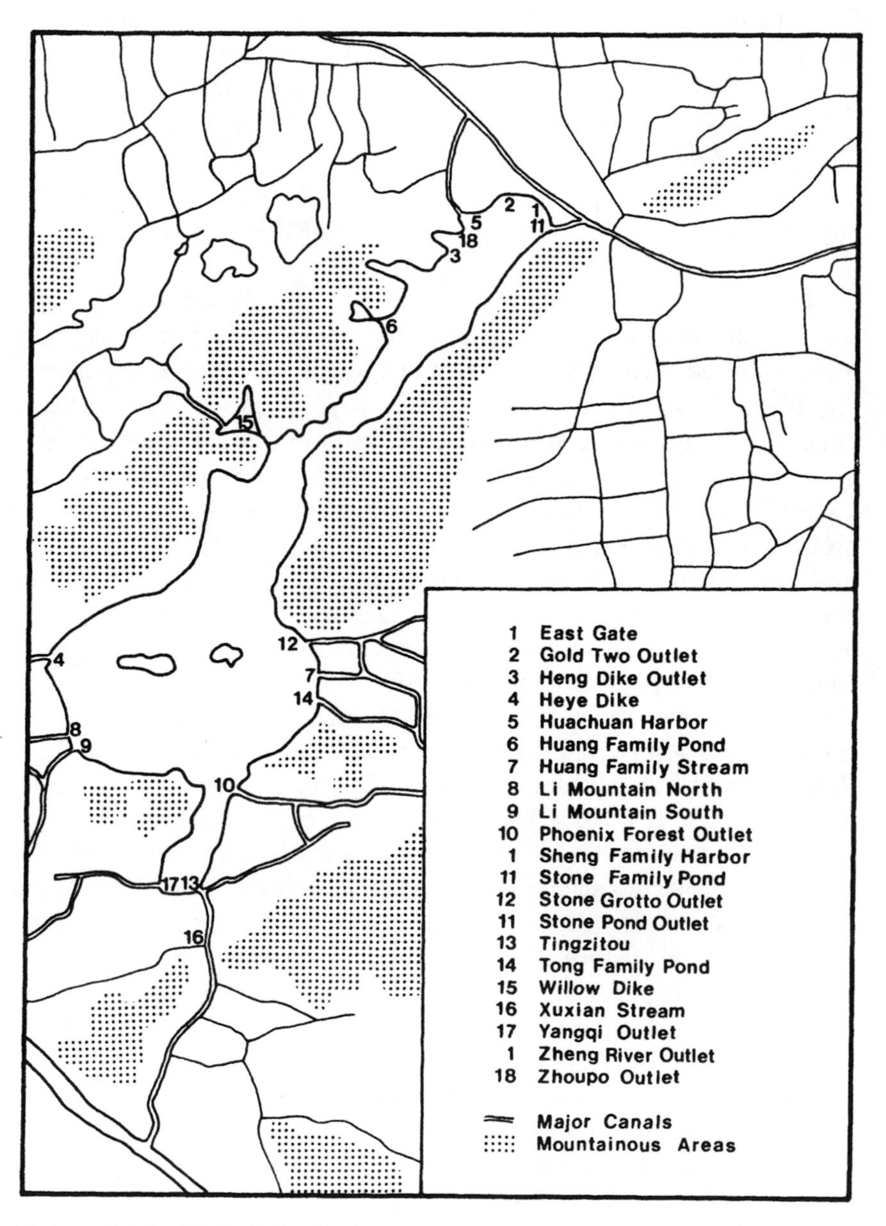

Map 4. Original Lake-Dike Outlets

Yuanming—were essentially definers of the lake, delineating and rectifying its use through regulations and protecting its limits against encroachments. Theirs was a difficult task, perhaps even more so than Yang's. Once the decision to build the lake had been made, Yang could depend on the excited commitment of the benefiting community to complete the task. To sustain over many years that commitment in the more mundane tasks of surveillance and maintenance was the challenge that faced later preservers of the lake.

Defending the Lake: The Courageous Stands of Zhao Shanqi

The allotment regulations set down by Yang in the early twelfth century have been lost, but whatever they were, by 1158 they were not being followed. In that year a drought brought on many disputes over the allotment of water. Disputes went beyond words, as the Xiang Lake community was torn by numerous bloody fights and by litigation prompted by personal attacks. The assistant magistrate Zhao Shanqi, assembled all dike chiefs and "upper household" heads to discuss the problem and decide on an equitable system of water usage.[14] The regulations, taking into account the altitude and placement of the irrigated land, set down six time periods for drainage outlets around the lake to be opened on a staggered basis. The acreage to be irrigated per village, the amount of water, and the amount of time the water could flow to each village's paddies were clearly specified. The lake could only be used for irrigation purposes from three days after the beginning of autumn (*liqiu*) in the lunar calendar (early to mid-August) until three days after the lunar calendar's White Dew (*bailu*) (mid- to late September).[15]

The size and construction of drainage outlets were prescribed: five Chinese feet wide (about 70 English inches) and three Chinese feet (about 42 inches) below the water surface. The sides were to be of cassia wood and the bottoms of stone to prevent the flowing water from tearing the outlet deeper and wider. Punishments were prescribed for violators. Anyone caught taking water from the lake out of the proper sequence for opening lake outlets would have his arms cut off; anyone found cutting his own drainage outlet in order to steal water would have his feet cut off.[16]

The sources do not detail how Zhao dealt with local figures whose anxiety for water had produced illegal drainage outlets cut into dikes or more ingenious ways of water pilferage. The Xiang Lake community acquiesced to Zhao's new regulations, though sources indicate that some were probably not satisfied either with the regulations for their native place or that equity had been reached in the apportionment of water. Zhao stood firm in his resolve to end irrigation contentions. In the end, as Mao Qiling simply stated, "There were none who dared dispute [Zhao]."[17]

A more serious threat to the lake's integrity was on the horizon: encroachment on the lake by people from outside the Xiang Lake community.[18] Less than two decades after the lake was completed in 1112, the Northern Song government was defeated in invasions by Jurchen tribes from Manchuria. In 1127, with the establishment at Hangzhou of the refugee Song court, Hangzhou became the imperial capital. The city's changed status transformed Xiaoshan county into an area of great strategic importance for the court. Permanent garrisons were established at Xixing on the Qiantang River across from Hangzhou, at three locations on Xiaoshan's Hangzhou Bay coast, and at the town of Fish Lake on the Qiantang River south of Xiang Lake. The number of soldiers and military guards stationed in the county were greater than anytime previously or anytime since.[19] With powerful government figures and bureaucrats less than ten miles away, Xiang Lake and other lakes in its environs became lures for various types of encroachers and reclaimers.

In 1167 three higher level officials colluded with some men around the lake to transform lake into land for private use. Two Hangzhou military commissioners, Li Xianzhong and Zhou Ren, took steps to become owners of lake land. Li wanted land in a beautiful natural setting in order to construct a country house. Zhou, who apparently already owned land in the area and was married to the daughter of a powerful area family, the Zhangs, simply instructed his tenants to encroach on the lake by planting a number of mu of rice.[20] In addition, area prefect Wang Qu conspired with a certain man near the lake to change lake to paddy land.[21]

When Zhao Shanqi became aware of these actions by officials, he took advantage of a previously planned trip to Hangzhou to approach the court about these matters. His petition reached officials at the highest level of government. Zhao met the vice president of the Board of Revenue and, after preparing a memorial on the matter, the prime minister himself. There he spoke vigorously in defense of the lake, stressing the plight of farmers should the lake be encroached upon. From Hangzhou, Zhao traveled to Shaoxing, the prefectural capital, using harsh and emphatic language to make his points with the offending prefect himself. Zhao becomes a symbol of the upright lower official, putting his own career in jeopardy for the sake of the local community, jawboning with the prime minister and prefect, and battling imperial commissioners and his bureaucratic superior out to further their own interests. To his credit, his courage seemed to have carried the day. Embodying in part the Chinese belief in the power of the written word, a signboard was posted at the lake forbidding the various encroachment schemes. But since there is no record that Li, Zhou, or Wang were reprimanded or penalized in any way, the "solution" seems one of form—to pacify Zhao and other accusers temporarily—rather than substance. Extremely important in the Chinese context, however, the formal "solution" provided a way for those in Hangzhou to express concern for the people of Xiang Lake, while the substance—failing to take substantive action against the indicted—gave evidence of the equation of political power at the capital.

Gu Zhong and the Crisis of Encroachment in the 1180s

When Magistrate Gu Zhong came to Xiaoshan in 1182, the lake had provided water to the area for seventy years and Zhao's regulations had been in effect for over two decades. Gu undertook a study of the whole area's water conservancy. Most of his findings related to Xiang Lake and circumstances around it. Gu found that there was indeed still a problem in the water allotment: while eight townships generally received enough, Xuxian to the southeast of the lake was receiving little. During the twenty years, some measures had been taken by the townships to rectify the situation, but the old regulations had not been changed to reflect that adjustment. Gu called together local elites, who quickly ratified the arrangement.[22]

But the question of water allotment was the least of Gu's difficulties. The first half of the decade of the 1180s was filled with disaster for the people of Xiaoshan county. Severe drought in 1180 and 1182 alternated with devastating floods in 1181 and 1185 to produce a farm populace fleeing elsewhere in desperation. Wives and children were sold. While fields lay fallow, the harvest of the dead was abundant. Gu described the situation as an unprecedented tragedy.[23] The government in Hangzhou provided general official relief grain and canceled tax payments for a year, action Gu attributed to the love of the Son of Heaven for his people and to his compassion for their fates.

Gu believed that the floods and drought came in large part from the widespread policy of reclaiming lakes on the Ning-Shao Plain. Little by little, people encroached on the sides of lakes, planting and harvesting rice, slowly filling them in. The lakes, thereby reduced in size, could not contain almost continuous spring rains that began in May and often overflowed from the sudden cloudbursts of summer.

At the crux of the problem was the relationship between population, land, and water. Water nourished the rice paddies, which in turn supported people. As the population increased, land per capita decreased and more land was sought; but as land was reclaimed and increased, water to nourish the rice paddies decreased. The area of Xiaoshan is subtropical, with an annual precipitation of 1,300 to 1,500 millimeters, generally sufficient for farmers' use. However, roughly between mid-July and the end of September, there is great heat with much evaporation. For late ripening, nonglutinous paddy rice, generally planted in June and harvested in October, natural rainfall was insufficient in the crucial part of the growing season; therefore, measurable water reserves for irrigation were needed.[24] In maintaining an ecological balance between land and water, however, the crucial variable was population, which increased markedly during the latter Northern Song and the Southern Song. In part the population swelled in the late 1120s from northern refugees following the victory of the Manchurian tribes over the government. In part the increase came from waves of northern emigration following such natural disasters as the flooding of the Yellow River in the late twelfth and early thirteenth centuries.[25]

The population of the seven counties of the Ning-Shao Plain passed 1.5 million by the mid-twelfth century, an estimated 600 percent increase from the third century. Such population pressure stimulated trends to reclaim lake land to produce more food. The historical record appears to verify Gu's analysis. The Tang dynasty (618–907), when the area had seen many lakes built or restored, had seen only six floods in the period from 648–831, an average of one every thirty years; in that time, only one drought had occurred. From 1134 to 1272, however, after years of lake reclamation, there were nineteen floods and eight droughts, one every five years.[26] The weather-related disasters of the early 1180s were clear indicators to Gu that the balance of water and land was askew and that lake encroachers were largely to blame.

Gu faced encroachment not only at Xiang Lake but at nearby White Horse, Shooting Star, and Zhan Family lakes as well. At all but the last, a wealthy civil official, an intendant named Zhang, turned out to be a central schemer. At Xiang Lake the problem came to light with petty violence among commoners.[27] Near the end of July 1184, a man named

Wang disclosed to authorities that Farmer Li and six others were occupying lake land and planting rice on it. Four days later, when Li and several other poachers met Wang's older brother on the road, they blocked his way and began to curse him, attacking him for his brother's disclosure. Taunts turned to violence. Grabbing a large stone as a weapon, they fell on him, cutting open his head and injuring his back. They dragged the bleeding Wang to Zhang's mansion on the northern edge of the lake. We do not know what happened there, but Wang probably faced hostile interrogation and treatment.

Another investigation about the same time led directly to Intendant Zhang and began to open for Gu the range of Zhang's flagrant actions. Farmers in the northern part of the lake accused another farmer of planting rice on a large area of lake land near the northern dike. Gu discovered that Zhang had hired the farmer to plant the land. Angered, Gu made Zhang's encroachment known to his superiors, who ordered Zhang to rectify the situation. Zhang's proxy poachers each received sentences of one hundred blows with a heavy wooden rod. Such penalties informed the local community that Gu was intent on dealing harshly with encroachers.

Even more than his encroachment, Zhang's construction of an imposing residence behind one of Xiang Lake's most important northern dike outlets reveals his almost incredible arrogance.[28] Xiang Lake lay to the south of the Grand Canal, while parts of two townships that received water from the lake were to the north.[29] Irrigation water traveled from the lake through the Zheng River outlet to the canal and from there through the Wanghu Bridge outlet to the townships. Zhang had built his home between the Zheng River outlet and the canal; behind the outlet he had constructed an earthen embankment strengthened with stones, creating a personal small lake in front of his mansion in which he kept boats and planted lotuses. Lake water, in fact, had not been able to pass through Zheng River outlet for over two decades, depriving farmers north of the canal. Yet the sources reveal not a single word of protest or petition for redress. This astounding fact points to Zhang's ability to cow and silence farmers who depended on the lake for irrigation. Apparently, only Gu's aggressive policies of lake restoration made possible the unleashing of the area farmers' complaints against Zhang and others. Zhang had further constructed a boathouse atop the Zheng River dike, the building of which damaged the body of the dike itself. After these misdeeds came to light in the wake of Gu's initial investigations, Gu personally directed fifty local men in the removal of the embankment, thus restoring the site destroyed by Intendant Zhang.

Encroachment at White Horse Lake, a smaller reservoir (about 450 acres) just over the mountains to the west of Xiang Lake, also involved Zhang.[30] Since 1173, he and another local power-holder had steadily turned the lake into paddy land and constructed residences that blocked irrigation flow. In late winter 1184, the prefect received a petition asking that the lake be dredged and restored, claiming that the unauthorized reclamation of the lake had produced successive years of drought, which was spawning starvation and vagrancy.[31] Authorities ordered the lake's restoration. Their specific instructions illustrate the political use of shame, often a potent instrument in Chinese culture.[32] Zhang, official though he was, was ordered to personally remove the soil that had filled the lake. The government further attempted to shame him by erecting alongside the lake a placard castigating Zhang and his co-reclaimer as members of elite families who brazenly assumed that their private affairs superseded the interests of anyone else.[33] Both Zhang's removal of the soil and the placard meant great loss of "face" for the guilty official.

Probably because of his widespread guilt in encroachment, the sources blame Intendant Zhang for enticing into reclamation activities Military Commissioner Zhou Ren (with whom Zhao Shanqi had already dealt). Zhang was apparently seen as some sort of an embodiment of vice whose alleged influence reached much farther than was actually possible. In this case, Zhou had been encroaching on lake land almost a decade before Zhang. In 1166, a year before he instructed his tenants to usurp land from Xiang Lake, Zhou had moved to take almost half of Shooting Star Lake, eight miles west of the county seat, an area landmark since the Han dynasty; he also took portions of Catalpa Lake and Dripping Melon Lake.[34] Zhou's wife, née Zhang, apparently with higher scruples than either her husband or blood relative and a remarkably independent woman for the time that saw the beginnings of the custom of foot-binding, asked the court that the lakes be returned to the people since they had existed originally for irrigation. Zhou had obviously been unable to intimidate his wife as he had the local populace. The court, dominated by men and protecting its own officials, did not act, on the grounds that none of the area's people had lodged complaints about it.

The natural disasters of the 1180s and the activities of Gu Zhong, however, prompted a number of men from western townships to accuse Intendant Zhang of destroying Shooting Star Lake. Military Commissioner Zhou was not impeached even though he had been the foremost encroacher on this lake for almost two decades. Such a glaring omission suggests important dynamics in Chinese social behavior. It is likely that

Zhou was seen by area inhabitants as too powerful a presence on the local scene to accuse directly. Silence before and acquiescence to authority was a cultural approach deeply embedded in Chinese society and politics (though, as with any cultural norm, exceptions occurred under various circumstances). Intendant Zhang, who had earlier silenced those behind the Zheng River outlet at Xiang Lake, was by this time seen as one whose power and authority had been deeply undercut by the public awareness of his misdeeds. In other words, it had become possible to accuse and attack Zhang without great fear of retaliation. Because of the connections between Zhang and Zhou, it also became possible to touch Zhou indirectly by accusing Zhang. Circumstances thus permitted maintaining the *form* of the power-status relationship (that is, commoners subservient to a sociopolitical superior—not accusing Zhou directly), but they also allowed the expression of dissatisfaction through indirection.

In a culture where personal standing and relationships rather than impersonal law were decisive in attaining social and political goals, the climate for redress depended on the status, motivation, and relationships of the people in key roles. If Zhang's position had not been undermined, we may conjecture that the Zhang-Zhou alliance could have better withstood accusations. More importantly and with more certainty, if Magistrate Gu had not waged such open campaigns against encroachers, it is unlikely that accusatory petitions seeking to end encroachment would ever have been sent. If Gu had not so aggressively defended the lakes, some of Zhang's unscrupulous actions would likely have been sustained because of Zhang's connections (*guanxi*) to Zhou. Gu was thus the central character in the restoration of the 1180s. His willingness to contend with higher officials makes it likely that he himself had higher official "connections," though sources do not disclose them.

Not sharing the hesitancy of the petitioners, Gu accused Commissioner Zhou directly, charging illicit involvement in reclamation. Zhou's wife, he averred, would never have requested the restoration of the lake if her husband had been acting with propriety. The proper policy, he argued, was lake conservation: if Zhou's wife's request were granted, the flood-drought problems of the county would be solved. When it became known that Gu would take up the request of the townships, Zhang, probably to forestall further formal charges of wrongdoing, hit upon the plan of petitioning for the restoration of the lake himself, which he did in December 1184.[35] It is notable that despite his culpability Zhang still held his official position as intendant. Though Gu had successfully challenged his actions, Zhang, probably on the basis of continued support from higher places, retained his post.

Finally, at Zhan Family Lake just west of Xiang Lake, Gu faced an encroacher who was a member of the imperial family itself.[36] The lake had been built by a strong lineage, the Zhan; however, the lineage fell on hard times, and members scattered, with the area's land falling into the hands of other families. During Gu Zhong's tenure as magistrate, a commoner presented some of the lake to an official, a relative of the imperial family. The presenting of land to officials was a common strategy of reclaimers and their accomplices; in this case, as in other similar instances, it is difficult to determine whether the presentation was initiated by the commoner or engineered by the official.

Gu was besieged with petitions on the Zhan Family Lake reclamation since the imperial relative had turned much of the small maple-surrounded lake to paddy and blocked the remainder from being used for irrigation. Petitions asked Gu, whom area people already saw as an opponent of reclaimers, to restore the lake and repair the dikes. Gu's investigation showed that the imperial relative had indeed obstructed the irrigation process. Unable to contend directly with the relative, Gu asked for funds to repair the dikes. His superiors responded that funds were unavailable since the boat on which the emperor often traveled via the Grand Canal to the imperial tombs near Shaoxing needed repair and thus required available funds. The emperor, his consorts, and relatives often used the Grand Canal for excursions to Xiaoshan and Shaoxing; on more than one occasion, empresses carried tablets from the court to be placed in local temples in the Xiang Lake area.[37] For the boat repair, the county provided more than four hundred strings of cash although only something more than three hundred had been requested. On Zhan Family Lake, by contrast, there was no action at all.

The original petitioner broadened his plea, asking that official usurpation of lake land in general be halted.[38] Hangzhou's response was that the imperial relative had purchased the paddy land but that the remainder of the lake was a reservoir. The imperial relative arrogantly refused to allow farmers with irrigation equipment access to the portion that even Hangzhou recognized as a reservoir. That refusal stimulated more petitions, which, with Gu's leadership, set in motion a long chain of events leading eventually to the lake's restoration.[39]

Gu congratulated himself on the restoration of the lakes, as well he might.[40] Within two years after his arrival he had succeeded, if not in breaking the arrogant stranglehold of some officials on the locality's resources, at least in controlling some of the more flagrant excesses. He also dealt with local minor wrongdoers in Xiang Lake—farmers who raised fish or planted lotuses to the disruption of irrigation.[41] In addition to meeting the encroachment threat directly, Gu also dealt successfully with the threat of wider disaster in the June 1185 floods. After heavy rains inundated the area, Gu personally traveled by boat to view the dikes, calling together local leaders to make dikes higher and to reinforce earth around the dike outlets.[42] He feared the breaking of the dikes not only because of resulting floods but also for the lack of irrigation water in the succeeding months should they wash out. To decrease water pressure by allowing the excess to drain into Hangzhou Bay, Gu opened the Zheng River outlet to the Grand Canal, possible only because he had been able to deal with Intendant Zhang's blockage of this outlet. He ordered that it be closed as soon as the rain stopped. The wisdom of Gu's care in maintaining the water in Xiang Lake was evidenced by the bumper crops in the area that year even though little rain fell after the floods of June.

Drawing the Color Line: The Work of Guo Yuanming

Three decades after Gu's magistracy, Guo Yuanming assumed the post.[43] The intervening years had seen people who lived at the foot of the mountains along the east and west shores of the lake construct some buildings on the lake's bank. Investigation by area elders confirmed these developments. The alleged encroachers argued that they had built not on lake land but on mountain land; the elders argued otherwise. Guo was called into the wrangle to settle the boundary issue. The sources picture him as uncertain and hesitant over how to delineate specifically the demarcation between the two types of land.

At that moment his fifteen-year-old son came forward with a suggestion: "The land is easy to distinguish: the land which is yellow is mountain land; the land which is greenish and light is lake land." This outspokenness of an adolescent in matters of public policy is as remarkable in Chinese cultural terms as the forceful assertion into political affairs of Zhou Ren's wife. Guo's son's suggestion, in contrast to that of Zhou Ren's wife, was adopted; while it might threaten the position of poor poachers, it obviously did not threaten the official male powers. Guo Yuanming reacted with enthusiasm; after dredging the lake, which caused no small uproar among those encroachers who had feigned innocence in their constructions, he proclaimed the line of soil color the boundary between the foot of the mountains and the body of the lake.

As one might expect, some sought to use the color line to their own advantage. Not much later one or more people dumped baskets of yellow soil into the water along the lake bank and set out to construct a dwelling. When Guo went to investigate, he saw the yellow of the mountain soil and was initially deceived. But the would-be encroacher was foiled by none other than Guo's son, who encouraged his father to return to the site and dig in the soil. When Guo followed his son's suggestion, the green was clearly visible. The dwelling was destroyed and the culprit dispatched to the army.

The Legacy of the Four

In the first century of Xiang Lake, then, four distinguished officials—Yang Shi, Zhao Shanqi, Gu Zhong, and Guo Yuanming—created and preserved the lake intact for the use of the irrigation community. The proximity of the imperial capital, with its plethora of grasping, self-interested officials, had made this period a particularly dangerous one for the integrity of the lake. In 1377, Magistrate Zhang Mou built a public temple on the edge of the lake to commemorate the accomplishments of the lake creator and preservers at spring and autumn sacrifices: it was called the Shrine of the Four Senior Officials.

Not only irrigators would have been willing to sacrifice there; many others benefited from the lake as well. Fishermen netted many varieties of fish: especially prized were the bream known for its jade green color and delicious delicate flavor, said by an early account to be "Yue's crown" (Yue, an ancient kingdom in the area, had given its name to the region), and the *dufu*, most abundant in the spring when the peach was in blossom, and especially delicious, it was said, when water in the lake rose after spring rains.[44] The lake teemed with delicious crabs. There were other water-products as well: two varieties of water chestnut; seeds of a plant akin to a water lily, used as coarse food and medicine and collected in autumn by the area's women; aquatic grasses eaten as vegetables; and, most famous of all, a type of lotus called the water-shield plant (*brasenia purpurea*), whose leaves and stems were a silky delicacy in soup and whose roots were used like wild rice by the poor and by many in time of famine.[45] The flavor of this edible plant from Xiang Lake was renowned. It was said that only the local people around the lake could distinguish it from the duckweed and other plants with which it grew intermixed; and it was averred that only the locals knew the secret method of washing and cooking it to make it soft and succulent.[46]

Poets, artists, and Chinese in general, who evidenced a deep love of nature's scenery, also owed debts to the Four. The scenery surrounding the lake prompted its name, according to Qian Zai, a Ming dynasty writer from Shaoxing, because it reminded someone of the landscape around a tributary of the Xiang River in Hunan province.[47] A few Song and early Yuan poems on the lake survive, and several point out the practical irrigative function of the lake: "One-tenth of an ounce of lake water is intimately related to the ten thousand mu."[48] But they describe primarily the lake's natural abundance and beauty: the luxuriance of the water-shield plant, the fragrant grasses, and the lotus flowers; the boats and oars of the fishermen; and the waterbirds—the crane in the shore grasses and the seagulls, ever-present reminders of the proximity of the bay. The domain of nature that overwhelmed Cai Panlong on his outing in 1271 is the world of the poets.[49] It is predominantly a world of the late afternoon, of dusk, when the blue haze and the slanting, weakening rays of sunlight bring a sense of loneliness, perhaps even of pathos, in the realization of the brevity of life compared to the lasting natural world of beauty into which this man-made lake had developed.

What remains notable about Yang, Zhao, Gu, and Guo, in light of the world of nature depicted by poets, is their humanity evidenced in the historical record. In contrast to Intendant Zhang (whose personal names we are never told) or Commissioner Zhou (whose only human quality seems to be his wife), images of the Four spring readily to mind: Yang bucking the trends of history to establish a lake when most were being reclaimed and dealing with landowners whose land, however unproductive, would thenceforth be submerged by water; Zhao, a minor official, gesturing, declaring, and going to extremes to convince his superiors to protect a lake for which he, through force of character, had set down regulations; Gu taking on his superiors who were trying to reclaim the lake, sincerely concerned for the populace to restore all area lakes, and personally visiting dikes endangered by flood; and Guo, human in his hesitancy and uncertainty, but with the enthusiasm and courage to accept the advice of his teenage son. Zhang Mou, the builder of their shrine, clearly saw that they were larger than life and that their legacy should be respected and perpetuated.

It is not surprising that Mao Qiling, the lake's seventeenth-century chronicler, was obviously disconcerted in reporting the subsequent history of the Shrine of the Four Senior Officials: "Afterward—I don't know when—the name was suddenly changed to the Shrine of the Two Senior Officials—Yang and Gu—and they were made the dukes of the Xiang Lake area. [Later] the name was distorted from Yang-Gu to Yang-Guo and the shrine was moved alongside the Pure Land Temple near the lake's Stone Pond Outlet."[50]

Like the Qiantang River currents and silt that obliterated the Raksha rock, the passage of time blotted out the accomplishments and differences of the Four. Gu's superhuman work of the 1180s became merged in the collective historical consciousness with the bumbling, though human, Guo. The efforts of the Four commemorated by the public shrine came to coexist with a Buddhist temple, a high irony, given considerable encroachment at area lakes by Buddhist monks in the late twelfth and early thirteenth centuries.[51] One is reminded of Cai Panlong's musings on Xiang Lake: "I have lost track of time. Images, pictures come and go—one forgotten as it is replaced by the next. I am both amazed and confused."

In the early fifteenth century a scholar named Huang Zong, concerned with the obliteration wrought by time, wrote a poem and commentary on the Pavilion for Viewing the River and the Lake, the views from which had so moved Cai Panlong.[52]

> Pavilion long since destroyed—but the name seems to suggest it still remains:
> The lake water with waves the color of the heavens, separated from the sound of the river.
>
> We hurry to find the foundation where our ancestors collected the arts.
> Though willing to carry on their traditions, we are ashamed
> Because posterity finds only wild grass and spring flowers in villages along the shore.
>
> The records show this pavilion was destroyed long before the Yong'le period (1403–1424). It is, in fact, sometimes confused by people with the One View Pavilion on Stone Grotto Mountain. Along the lake shore at Tingzitou village, there is a bit of wasteland where there are many pieces and fragments of broken stone—perhaps this was its site.

II

VIEW

Chenxi Garden, Late Fifteenth Century

CHAPTER

Obligation and Death, 1378–1500

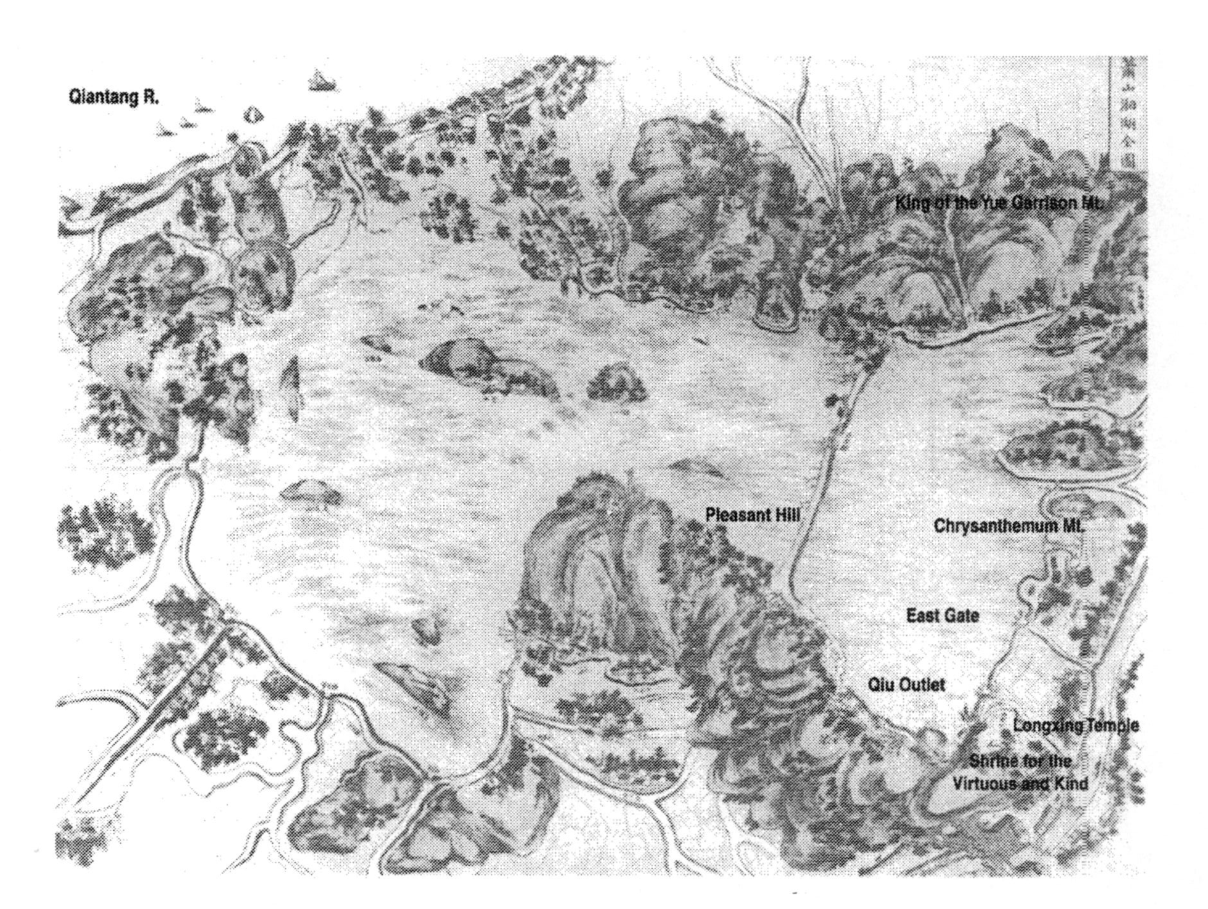

Map 5. Xiang Lake Sites, Fifteenth Century

VIEW: *Chenxi Garden, Late Fifteenth Century*

It was not a notably large garden. But if a Chinese garden was meant to replicate natural scenes by incorporating elements within—pavilion, rocks, bamboo—with vistas beyond the garden, then it was undoubtedly a popular site for serious viewers of scenic beauty.[1] Located on the lake's northern shore, it would have allowed onlookers from its almost obligatory pavilion many prospects of lake, mountains, heavens, and natural plant and animal life. Though the county seat with its ruined ancient wall could be seen in the distance through the clumps of bamboo, the garden was far enough into the country to allow those from town to escape, however briefly, into the world of nature.[2]

No record remains of the arrangement of the garden, though from accounts of other gardens in the area we may make reasonably accurate assumptions about its component features. In all four seasons the garden should have offered the cultivated viewer pleasing sights.

Winter in Xiaoshan often ignores the subtropical categorization of the lake and its environs. In a later period, Cai Mingheng described the piercingly cold north wind driving snowflakes into his face as he passed the nearby garden at the base of North Trunk Mountain.[3] During such cold spells, the garden should have continued to exude signs of life. The "three friends of winter"—bamboo, pine, and chrysanthemum—were thus essential components.[4] So crucial were they that for the Chinese they have become significant symbols: the durable and resilient bamboo, whose

Figure 1. Chrysanthemum Mountain (1986)

segmented growth evinced a sense of standards and limit, symbol for the Chinese gentleman-official; the pine, green in winter with gnarled bark against the slate sky or wind-driven snow, symbol of the Confucian scholar's persistence and tenacity in the face of oppression and hard times; and the chrysanthemum, which flowers when others have long died, symbol of endurance amid the vicissitudes of season and condition. Less than half a mile to the west of Chenxi Garden rose Chrysanthemum Mountain, and it is likely that on an early winter day, perhaps framed by pine or bamboo in the garden, the golds, lavenders, and coppers of the chrysanthemums that covered the mountain could be seen against a blanket of snow (figure 1).[5] The snowfall would have made stark the line between dark lake and white mountain, certainly a more distinct and immovable demarcation than Guo Yuanming's color line could ever have been.

While bamboo and pine retained their rich greens throughout the year, the first sign of spring came with the almost miraculous appearance of delicate red-and-white plum blossoms on naked, rough branches. Able to endure the still wintry blasts of late winter, these blossoms clothed the mountains around Xiang Lake in February and March. From the garden they appeared on the mountains to the west and south in natural abundance. The creator of the garden would almost certainly have cultivated one or more plums, which shared with the pine the symbolism of hardiness amid bleak hostility. Each year it seemed that the appearance of plum blossoms was in part simply the signal for other, more colorfully and fragrantly flamboyant flowers to spring forth: wintersweet blossoms of intense jasmine fragrance; the magnolia's creamy pink-white flowers; the crab apple's mildly fragrant clusters of tiny red, pink, and white blossoms; and by May the peony's wealth of petals in yellows, purples, and reds.

It was almost certainly the peonies that Magistrate Wu Shu noted on the northern shore of the lake and on Chrysanthemum Mountain one clear spring evening in 1480.[6] Taking a sedan chair from the Shrine for the Virtuous and Kind, established for the memory of Yang Shi of the Song and Wei Ji of the early Ming, who had worked to uphold the lake, Wu passed the Longxing Temple on his way to the first drainage channel from the lake, the Stone Pond Outlet. Both shrine and temple were clearly visible from Chenxi Garden and, to a scholar familiar with the area, were reminders of the transcendent importance of water and defense for the area's survival. Longxing Temple had been built by a general in the 930s, had been destroyed in the bloody turmoil at the end of the Mongol's Yuan dynasty, and had been rebuilt only in the early fifteenth century.[7] Because spring rains had flooded the path around the lake, Wu traveled by boat after reaching the Stone Pond Outlet. Although he passed Chenxi Garden, it did not elicit his comments; perhaps he was more interested in the silhouette of bamboo and cassia against the setting sun, perhaps the wine he was drinking had combined with the rocking boat to produce a groggy drowsiness. In any case, he headed for a county landmark, the King of Yue Garrison Mountain, for the night.

As one journeyed west from the county seat, Chenxi Garden was located at the East Gate, the second drainage opening along the dike. In mid-August it began to be opened for periods of over twenty-one hours during the third of the lake's staggered openings. By then the lotus blossoms in the garden pond had become great disks of color in the air above the lotus leaves; the daylilies and irises along the pool's edge were swatches of purples and oranges and blues against the dark green azalea and wintersweet leaves; the cicadas throbbed in the shimmering humidity; and the whine of the mosquitoes, omnipresent in this water-filled world, could be halted only temporarily by the vigorous fan waving of a determined scholar as he absorbed the summer scenery.

Viewed from the garden, the lake was the realm of fishermen, irrigators, mud-dredgers, and, especially in late afternoon and evening, those escaping the heat of town and the rigors of work by sailing or rowing with friends and, more than likely, by drinking wine. In gardens thoughtful Chinese sought release from the irritations, frustrations, discord, and cares of life. Magistrate Wu Shu, in his 1480 journey around the lake, did note a garden east of Xiang Lake with pavilion and pond amid a flourishing of flowers, great bamboos, birds, and a herd of deer: viewing that garden, he recorded, removed his anxiety and the cares of his heart, and he was even moved to write a song expressing his happiness. And what was Xiang Lake except a garden writ large? Its mountains, its waters, its foliage and flowers, its clouds and mists, its ability to bring relaxation and closeness to nature—all this drew people to the lake, eliciting poetry and essays extolling its beauty.

About the time the period of irrigation ended in September, the first signs of autumn began to appear in the garden.[8] The catalpa, which provided the wood for imperial coffins, began to turn color, a certain signal, people said, of the approach of the Mid-Autumn Festival. Immediately before and during the rice harvest, the view of the lake from the garden was most majestic: the summer's heat and its lake haze dissipated; the days were cooler and clearer; the lake's water shone a deeper blue to match the "kingfisher blue" of the heavens.[9] In days of autumn brightness, the cinnabarine maples on the mountains around the lake, and perhaps juxtaposed in the garden with the pine and bamboo, vaunted the splendid contrasts of nature. But if autumn brought splendor, it also brought death; with the harvest came the end of the growing season. The maples dropped their blood-red leaves to stand stark against the sky. The *wutong*, it was said, dropped one leaf a day to mark the passage of autumn.[10] The lake's harvest of the luxuriant water-shield plant was snuffed out when the first frost descended.[11] And the lotus, regal in the summer amid floating duckweed in the lake and garden pond, turned brown at the first cold weather, burned by the frost into a contortion of tendrils, petals, and leaves.[12]

Like the accomplishments of the Four Senior Officials and like the fate of the Pavilion for Viewing the River and the Lake, within several centuries Chenxi Garden was no more. The beauty of its human-planned and created scenery—tree, flower, rock, and pond—had been overwhelmed and destroyed by the natural growth of a bamboo forest, which in its luxuriance had also eliminated all open vistas of the lake.[13] Like Wu Shu, who passed by without comment, one might assume that this scenic garden view simply depicts the seasonal flora of Xiang Lake; but it is more important in the story of Xiang Lake as the setting of an extraordinary, violent episode that occurred one early autumn day in the year 1500.

CHAPTER: *Obligation and Death, 1378–1500*

Historical sources have not provided information about the reactions of lake viewers Cai Panlong and Sun Dehe to the advancing Mongol armies late in the decade of the 1270s. Conscientious young men in Xiaoshan reacted in different ways to the conquest by brutal outsiders. He Zongju, in his mid-thirties and associated with the Hangzhou court, fled in 1275 to Xiaoshan to live as a hermit, unable to serve a new lord.[1] Ding Shen, a brilliant twenty-three-year-old scholar, out of loyalty to the Song emperor threw himself to his death from a cliff when the Mongols crossed the Qiantang River: "Three centuries of the ancestral temples and altars to the spirits of the land destroyed in one day—how can I go on?"[2] Not all Xiaoshan men, of course, dropped out of society, either figuratively or literally. While the emphasis on loyalty to one's master or lord does not have the same unconditional force in Chinese tradition as the vassal-to-lord loyalty in feudal Japan, the obligation (and loyalty) one owes to one's superior—be he ancestor, father, teacher, or emperor—is grounded deeply in the classical culture.[3]

Given the cultural norms, the individual must decide, for example, if allegiance to particular ideals or a new regime should supersede other obligations to person or regime. Li Di, the Qin general whose spirit inhabited North Trunk Mountain, forsook the Qin regime to follow Xiang Yu: he was honored for his death in the service of Xiang Yu. A more famous general, Yue Fei of the early twelfth century, insisted on opposing the advance of northern tribes into the territory of the Southern Song even when his emperor, supported by a chief minister, refused to sanction that action. Eventually executed on a trumped-up charge of treason, he was posthumously canonized as loyal (*zhong*)—to ideals that transcended allegiance to individuals.[4] The Chinese had many varied, and acceptable, models in this regard.

Time obliterates not only natural gardens and man-made pavilions but also memories. For many Chinese, the political legitimacy of a regime generally seemed to be strengthened the longer it existed: even a regime as harsh and discriminatory to the southern Chinese as the Mongol held the Mandate of Heaven. Within less than a century, with the Mongol regime in disintegration and the rise of regional would-be emperors, county records make supporters of the Mongols seem positive forces of virtue when compared to their opponents, the future founders of the Ming dynasty.

In the beginning it seemed that Li Di's spirit might protect Xiaoshan from bloody warfare. Xiaoshan was, in the words of one commentator, the "throat" through which Eastern Zhejiang (*Zhedong*), with its flourishing trade, salt production, rice paddies, and sericulture, could be entered from Hangzhou.[5] In 1352, Fang Guozhen, an independent military leader who supported his ventures by raids on coastal towns, shipping, and government granaries in the present-day provinces of Jiangsu and Zhejiang, made ready to cross the river into Xiaoshan.[6] Zhao Cheng, a county official known for his public spirit, ascended North Trunk Mountain.[7] The results of his supplication were as spectacularly dramatic as the case of Fang La over two centuries earlier. In the middle of the night supernatural lights appeared on the Xiaoshan bank of the river, seeming to Fang Guozhen and his troops to be the lanterns of encamped soldiers. The apparition again proved efficacious: Fang did not cross.

Seven years later, however, it was a different story. In February 1359, the forces of Zhu Yuanzhang, who would establish the Ming, took Zhuji county to the south of Shaoxing.[8] In March his leading general planned an attack on Qianqing, nine miles southeast of the Xiaoshan county seat. The attack involved the purposeful destruction of over 235 yards of dike and caused considerable flooding with loss of property and life in both Xiaoshan and Shaoxing counties. It appears that Li Di's magic could check enemies only from across the Qiantang. What followed was high irony, given the county's sense of Li Di's place in its defense. In May a Ming general launched a devastating attack from Zhuji directly on Xiaoshan. When Yuan loyalists could not hold out, the Ming army plundered and burned the countryside: "Smoke darkened the sky" as Yuan loyalists fled across the river to the west. Ensconced in the county, the Ming army chose its headquarters, North Trunk Mountain, the very home of Li's spirit. There was a temporary reprieve from the Ming rule of Xiaoshan when Yuan generals brought together "righteous" soldiers on more than three hundred boats at Fish Lake. Ming forces retreated for a time, taking booty from the area. But by 1368 all of China came under Ming control.

We are not told of the plights of farmers and fishermen around Xiang Lake either in the victory or in the defeat of the Mongols. Certainly the opening of the dikes and the smoke-blackened sky suggest that the lives of the area's people were almost negligible to those who fought. Nor do we know if any of the lake community knew or cared who was fighting whom and for what. Endlessly repeating cycles of planting and harvest, punctuated by birth, marriage, death, and festival days, formed the rhythm of life. Such natural disasters as the tremendous wind and rain storm in the spring of 1333 that uprooted trees, destroyed crops and houses, and caused an unknown number of casualties only added to the sense of being a small part of an awesome natural world.[9]

Xiang Lake: Late Yuan, Early Ming

The situation at Xiang Lake had deteriorated rapidly during Mongol rule. By the late 1340s, much of its shallower sections were clogged with weeds, and, as a whole, water was insufficient.[10] Paddies around the lake became barren, and hungry people were ready to join bands of thieves that reportedly were forming in the mountains. A minor struggle occurred in the years 1338–1341 between two county officials, the magistrate and the chief judicial official: the magistrate, seeing the deteriorated shape of the lake, wanted its complete destruction, while the judge insisted on its preservation despite its clogged, encroached-upon state.[11] Though the judge won the dispute, at the end of the decade the lake's condition had not much improved. In 1349 a new county judicial official championed the lake, paying special attention to dredging lake areas and building up leaking and unrepaired dikes.

The political instability and the military exigencies of the era also took their toll on the lake. In addition to the fighting of the 1350s and 1360s, which damaged or destroyed lake dikes, drainage channels, water gates, and embankments, one magistrate actually promoted a policy of reclamation for the benefit of area residents. In the battles for control of territory to the southwest of the county, military on both sides demanded that county rice be shipped to the troops, leaving an already desperately hungry populace with no grain for themselves. The magistrate benevolently opened government granaries to feed the hungry and relaxed prohibitions against Xiang Lake reclamation.[12]

A new dynasty, like a new year, meant a fresh start. Nine years after its beginning, Magistrate Zhang Mou, who had the Shrine of the Four Senior Officials constructed, worked to reestablish the lake as a functioning irrigative reservoir. It had been two centuries since Gu Zhong's work. After his visit to the lake, Zhang described the crumbled stele with the regulations in dust, the private drainage outlets that had been cut, and the theft of water. He searched for the old records to discover which drainage openings were legal, but it was, he tells us, a vexatious task. He clearly saw himself as carrying on the work of the Four; his goal was to restore the system by reengraving the regulations in stone. By restoring the lake, Zhang was fulfilling an obligation not only to the people of the Xiang Lake irrigation community but also to the memories and records of Yang, Zhao, Gu, and Guo. He was also thereby placing himself in a line of protectors of the lake, marking himself as an upholder of the tradition begun by the Four. In a real sense he was establishing himself in an ancestral line, not of patriarchy but of role and accomplishment.

Buying the highest quality stone available, as the symbol of a new start, Zhang had the regulations carved and placed before the right column of the county hall in the county seat.[13] He finished the task in early October 1378, shortly after the Double Nine Festival, a day when people bought meat pies with little paper flags of different colors, held large banquets, or picnicked among the chrysanthemums in the mountains.Chrysanthemum and dogwood petals were traditionally placed on cups of wine as charms to eliminate evil powers. One can imagine Zhang's hoping that, with the new start he had given systematic irrigation at Xiang Lake, the Double Nine floral amulets would do their part to preserve the lake from harmful influences.[14]

The Work of Wei Ji

Wei Ji, who was honored with Yang Shi, the creator of Xiang Lake, in the building of the Shrine for the Virtuous and Kind in 1473, was of a different mold from Yang, Zhao, Gu, Guo, or Zhang. While the others were officials who served the area through their posts in the imperial bureaucracy, Wei was a native of Xiaoshan who in his retirement from a long line of distinguished positions made important contributions to his county's water conservancy. Such a career pattern was common for Chinese officials and points to the great significance of birthplace for Chinese in general. One felt a strong sense of the significance of one's native place as the home of familial ancestors; a strong social tie to others sharing that native place—whether village, county, prefecture, or even province; and an obligation to further the well-being and standing of that area vis-à-vis other areas.

Wei was born in 1374, three years before Zhang began his effort at historical restoration. He passed the provincial-level civil service examination at age thirty-one, thereby gaining for himself the second-level civil service degree, the *juren*. Whether he failed to pass the examination for the highest degree, the *jinshi* (at the metropolitan and palace examination), we are not told, although his quick appointment to serve at a prefectural school in Jiangsu province suggests, in light of early Ming practice, that this may have been the case.[15] He built a fine reputation at the school, serving for twelve years and helping during that time in the preparation of the Great Encyclopedia of the emperor Yong'le. He served in several posts in the Court of Imperial Sacrifices and was vice minister in the Board of Personnel and in the Board of Rites. During his years at court he also directed the metropolitan examination three times and the Jiangsu provincial examination twice. He ended his official career as Minister of Personnel of Nanjing in 1451, when he retired at age seventy-seven. During his retirement, which lasted another twenty years (he died at ninety-seven, the oldest Ming dynasty official[16]), he wrote at least ten poems on Xiang Lake and, more significant for his contemporaries, contributed to the lake's preservation.

Wei's poetic legacy is remarkable for its content and tone, especially when compared to the rather large body of Xiang Lake poetry.[17] Much of Chinese poetry is haunted by a sense of the brevity of life, of time running out, of the vanity or futility of human works. It is laden with images of autumn and evening, overwhelmed by an almost painful sense of endings. From a near octogenarian we would perhaps expect somber dirges, especially when we discover that seven of the ten poems he left are descriptions of trips to or past his planned gravesite. A clue to Wei's disposition and outlook comes from the name he gave the gravesite: Pleasant Hill.[18]

Wei's poems on Pleasant Hill generally take place in early morning or midday; despite allusions to and comments on retirement and death, their themes are life and the bounties of nature. In Wei's poetry, Xiang Lake indeed becomes a garden, providing varied views through the four seasons. Four poems are dated according to time of year: mid-August, late September, December, and April. At dawn in mid-August in 1453, Wei set sail for Pleasant Hill, recording the scenery of the lake. The tone is peaceful, bountiful, contented.

> There are ripples on the still lake.
> I take advantage of the dawn wind, let it
> push the sail to wherever it will.
> The glint of the sunlight is like
> gold flowers in the lake.
> In the heart of the waves are the
> strong-stemmed water-plants.
> Overhead are the mists; on the rocks are
> pines; cranes swoop over the water.
> Farmers cease from their labors and come
> quickly to welcome [me] . . . .

The poem of late September mentions the seasonal chrysanthemums counterposed to the jade green trees, a faraway shrine on the lake's shore, and a monk's yellow-thatched hut. Wei's treatment of human activity is remarkable: the world of nature is for human use and profit—a contrast to earlier (and later) descriptions that render humanity inconsequential amid the awesome power of nature. In the September poem, Wei talks of the farm family whose "sweet-smelling rice has already borne fruit" and of the woodcutter coming out of the valley who stops to loosen his load.[19]

Although Wei's descriptive December poem is filled with thoughts on his own situation, he paints a picture of purity and abundance—of snow falling onto his head and into the lake while streams cascade down mountains in the distance.[20] Only in his poem of April—ironically in springtime—does Wei turn to the more somber themes: the futile plans of men to direct their world, the ending of the day, the drinking of wine. But even then, it is a picture of nature's bounty with the mention of the lake filled with water, stocked with fish, and his passing the shrine to Yang Shi, the lake's creator.[21] Like the Chinese gentleman who viewed the garden from a pavilion, Wei from his boat sketched a world of the green willows growing near lake dikes, the smell of lotuses, the splash of oars, the "beautiful water and wood"—Xiang Lake infused with his own life-filled vision.[22] Nature's bounty was Wei's frequent theme. In a poem written in the summer of 1458, Wei juxtaposes the barrenness of drought with the bounty following use of the lake water for irrigation, significantly linking lake water to the happy shouts of farmers growing the grain that should lead to a successful harvest.[23]

Finally, in his "A Song about Xiang Lake," Wei sets forth his awe at the legacy from the past and his fears for the future:

> One hundred *li* around; vast areas of sand beaches.
> The bequest of Tortoise Mountain [Yang Shi]—who can ever forget this?
> The water, stored to irrigate, can contain the water from a thousand mountain torrents
> To flow to nine townships at time of drought.
> Farmers take edible water-plants and lotus to sell at markets: many seek to taste the water-shield and other plants.
> County leaders should stop treating the lake lightly: they should look to the dikes and manage carefully.[24]

Even before his return from official position and involvement in specific local water conservancy problems, Wei had shown a serious interest in water issues, especially lake management. In a memorial drafted in the summer of 1440, he had written that in the first years of the dynasty, each administrative state division (*fu, zhou, xian*) had dredged lakes and built dikes to ward off flood and drought.[25] But, he said, later officials failed to continue such vigilance, deputing people in lackadaisical fashion to see after the facilities and failing to question them about deteriorating situations. Local bullies and devious people assumed leadership of the dike building in order to set up their own diked areas where they could plant lotuses (all parts of which could be sold for medicine or food) or raise fish (setting basket traps for them on the dikes). Little by little the lakes became clogged, and the water available for irrigation decreased, thus harming the county by limiting food supply.

The court had responded to Wei's memorial with a prohibition on the building of private dikes by officials and calls for renovation of the lake lands. Punishment for transgressors included the possibility of jail, placement in a cangue and fetters, or assignment to an army unit on the northern frontier.

Though Wei did not necessarily have Xiang Lake specifically in mind in 1440, the history of the lake in the early Ming was punctuated with occasional attempts by area residents to grab lake land for themselves. At the turn of the fifteenth century, a tenant farmer named Su farmed alluvial soil built up on the river bank (apparently in the mouth of the Qiantang).[26] When the soil, termed official land (*guan tian*), collapsed into the unpredictable current, Su reported to officals that he would reclaim a similar acreage in lake land near the river to make up for his collapsed land. Officials took no action against Su: an illustration of the lackadaisical officials described by Wei. A few years later, others followed Su's lead; the government issued perfunctory prohibitions but apparently took no firm action to restore the lake.[27] Finally, during Wei's retirement, the situation had reached the point where area elders petitioned the county seat to clear away households that had reclaimed the lake and to figure an appropriate fine of grain per mu of reclaimed land to be placed into an official fund for relief purposes. Mao Qiling's account of the lake notes that even then the encroachment continued.[28]

Wei's first involvement in Xiaoshan water issues, however, came as a result of tremendous floods in June 1456. Heavy rains to the southwest in Quzhou and Yanzhou raised the Qiantang River, which flooded over Xiaoshan from the south, while the Little West River flooded from the east. Eighty-two years old, Wei took on the responsibility of managing emergency flood control, even carrying baskets full of stones to bolster dikes along the river to the east and southeast of Xiang Lake.[29] Though the flood damage was great, Wei's actions helped in reducing the disaster. Another serious flood in 1461 prompted Wei to undertake a general reconstruction of dikes and sluicegates.

The Excavation of Qiyan Mountain

During this period of Wei's reconstruction program, the government sponsored a momentous water control project about three miles southeast of Xiang Lake's most southern point. It had dramatic implications for some in the lake's irrigation community, far surpassing in importance the bit-by-bit encroachment irritations of petty reclaimers.

Before the period of the Tianshun reign (1457–1465), the Puyang River flowed from the southern counties of Yiwu, Pujiang, Dongyang, and Zhuji. Forming the boundary between Xiaoshan and Shanyin counties, it flowed past Maoshan, became a part of the Little West River, and flowed out to the sea at the Three River Outlet (*Sanjiang kou*). When heavy rains fell over the southern mountains, the narrow riverbed frequently could not contain the rushing river, which flowed over dikes, flooding the low-lying land in Xiaoshan and Shanyin counties. To try to solve this perennial flooding problem, Shaoxing prefects Peng Yi and Dai Hu made a study of the topography of the area and decided on a solution.

Less than a mile to the northwest of Maoshan was Qiyan Mountain; Peng and Dai decided to excavate a passageway through the mountain to rechannel most of the water of the Puyang to the northwest, where it would join the Qiantang River. In addition to relieving the pressure of excess water by diversion, Peng and Dai built an embankment and sluicegate at Maoshan to allow some regulation of the flow of water through low-lying areas to the sea (see map 6).[30]

This project, successfully completed by 1465, had two major effects on Xiang Lake. Most directly, it increased the risk of flooding from the south. When heavy rains fell on the central Zhejiang mountains, the Qiantang often rose to flood stage; after the project was completed, floodwaters from the Puyang River were added to the flooded channel of the Qiantang River. In effect, the project made flooding in northern low-lying areas near the sea less likely while increasing the risk in the Xiang Lake area. This new situation meant that Xiang Lake's viability as a reservoir depended on vigilant maintenance of the West River dike between Linpu and Xixing: breaches in the dike at time of flooding could destroy drainage ditches, lake dikes, and sluicegates and could deposit sediment harmful for irrigation.

A second major result of the excavation of Qiyan Mountain was that the flow of the Puyang River on its path to converging with the Qiantang cut across two townships, Anyang and Xuxian, that used lake water. This new development, placing some territory south of the river, permanently separated 80 percent of the land in these two townships from the lake.[31] There is no record of how the people affected by this drastic development reacted, nor is there even evidence that Peng and Dai were aware of all the effects of their water control project on the area. The irony of the situation is readily apparent: Xiang Lake could have a distinguished array of upholders of the lake—Yang, Zhao, Gu, Guo, Zhang, Wei—whose energy and diligence preserved the lake as reservoir. But government decisions to resolve another water control problem, seeming perhaps as arbitrary to the Xiang Lake community as nature's might itself, placed the effective long-term functioning of the lake in jeopardy and cut off all access to the lake water for some. The Qiyan Mountain project in the mid-fifteenth century began a long process of the diminution of the Xiang Lake community of irrigators and set up a permanent threat to the lake's effective functioning.

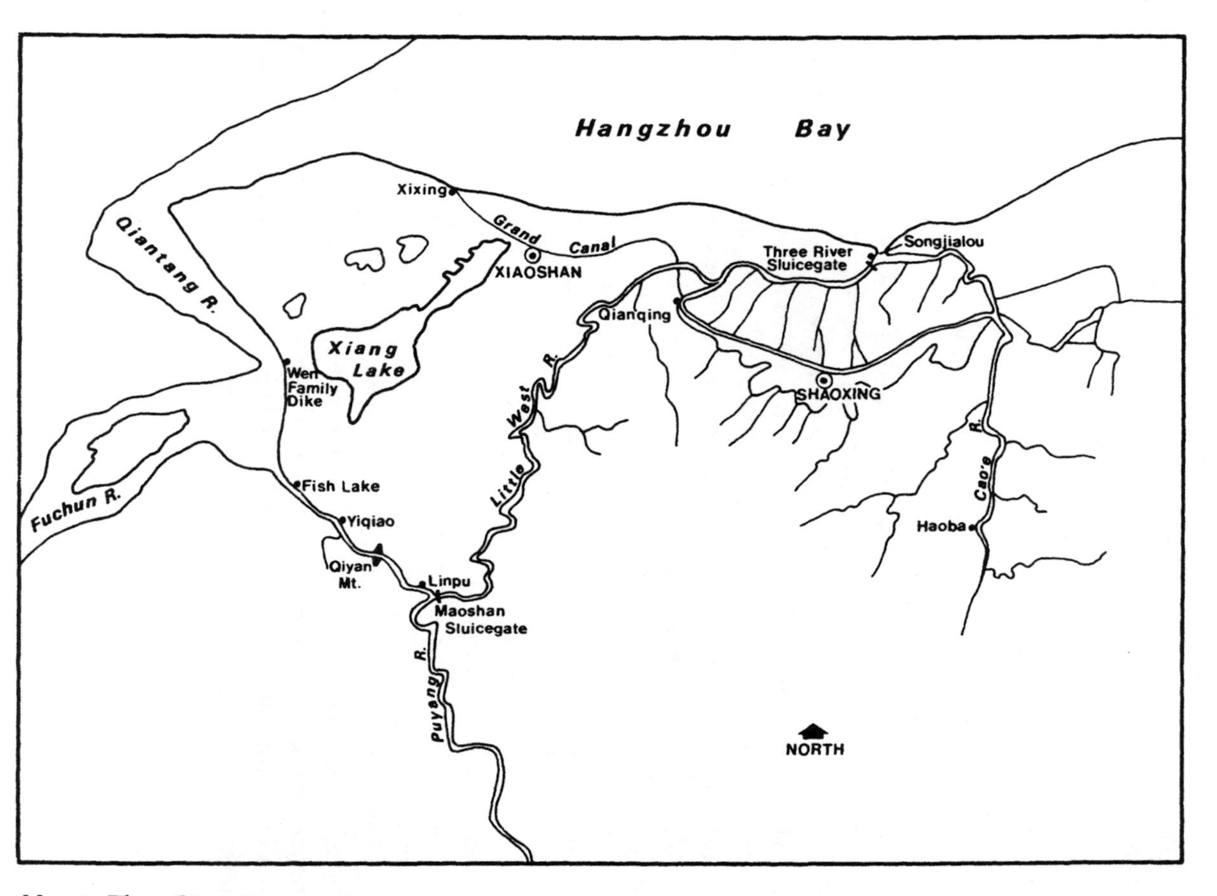

Map 6. Three-River Drainage System

The Affair of Censor He

Increasingly serious for the viability of the lake as reservoir was the level of encroachment reached around the lake by the 1460s. In most cases, encroachers simply planted rice or lotuses in a small area near the shore that had grown shallow over time. In some cases sheep were pastured or willows planted on dikes. Both practices eroded the dikes, making leakage or outright breaks more likely. Small-scale farmers took relatively small amounts. But taken together, the amount of the lake destroyed by encroachment was 7,318 mu (over 1,100 acres), almost 20 percent of the original size of the lake.[32]

Wei was determined to right the situation and restore prosperity through a major repair of dikes, sluicegates, and embankments and the dredging of encroached-upon land. In all this, he was remarkably successful. He had to deal with two relatively new families involved in encroachment, the Sun and Wu lineages. One of the Sun lineage, married to a Wu woman, secretly encroached on the lake to benefit himself and his wife's family. How active any of the Wu lineage was in this scheme we are not told, but Wei Ji was able to make known the Wu's encroachment and restore all their land to lake. He was not, for whatever reason, able to restore all of the land seized by Sun, who seemed the main culprit. The Sun threat could not, however, have seemed that ominous at the time: his personal name was not even recorded in the sources. Yet Wei was concerned, and before his death in October 1471, he charged his protégé He Shunbin with clearing the lake, declaring, "Whoever clears this lake is a man of virtue (*junzi*)."[33]

Through such a declaration, He Shunbin was charged with two obligations, one to his teacher and the other, through the Confucian import of the words of the charge, to be the servant of the people. We do not know the nature of Wei's and He's relationship. Shunbin received his juren degree in 1468 and his jinshi degree in 1469; it is possible that Wei tutored him. One of a long line of distinguished officials, He became a censor in Nanjing and in the Huguang circuit. Personally upright, he incurred the enmity of obstinate and undutiful officials, who, through ties to the court, were able to have him exiled to a garrison in Guangxi province.

During He's exile, the encroachment situation at Xiang Lake worsened dramatically. Leaders of the Sun lineage, adopting a strategy of intermarrying with the Wu lineage, had embarked on a major campaign to usurp lake land for their own uses. In the Chenghua reign (1465–1488), Sun Quan and his in-law Wu Can occupied 1,231 mu (over 185 acres), turning it into paddy land. Furthermore, even before the arrival of magistrate Zou Lu in the county in 1496, they had bribed him to remain silent about their betrayal of the lake. In contrast to the Song period pattern of encroachment (wealthy officials hiring local lackeys to extend their personal estates), the mid-fifteenth century saw the leaders of strong local lineages seducing local officials into silence and connivance.

When Censor He was pardoned for his offense, he returned to Xiaoshan; apprised of the reclamation by Sun and Wu and remembering the charge of his mentor, Wei Ji, he set out to restore the lake by eliminating the control of the lineages.[34] When he sent a memorial to the throne to rectify the situation, he was not yet aware of the involvement of Magistrate Zou. We know little about Zou Lu. He had taken the magistracy of Xiaoshan after serving in a military garrison on the western frontier. He was an active magistrate who saw after the needs of the county; in 1498, the year of He's return, Zou had sponsored the reconstruction of the Xixing town lighthouse.[35]

We do know that he took umbrage at He's intrusion into the affairs of Xiaoshan county. Even if Zou had not been colluding with the usurpers, one can imagine his irritation with the native son, so recently returned from exile by the throne, pointing out Zou's inadequacies. The actions of He were certainly intolerable to Sun Quan, who reportedly dispatched more presents to Zou; whether that was necessary to firm up Zou's resolve is not known. When Zou made known his intention to oppose Censor He's involvement, He reacted by attacking Zou. The animosity between the two men escalated rapidly. The official *Ming History* reported that Zou was "coveteous, cruel, crafty, and ruthless." Mao Qiling, more to the point, recorded that Zou began to feel a great hatred for He, who must have seemed unbearably self-righteous, especially in the context of the imputation of wrongdoing that had led to his exile.

Zou hit upon a plan to discredit He, using the exile itself as the main weapon. Zou ordered Sun Quan quietly to spread the word that Censor He had illegally returned to Xiaoshan without an official pardon. Because he therefore had no official seal with which to send the memorial, He had been forced to steal one. His memorial on the lake was thus fraudulent. Unable to argue on matters of substance, Zou resorted to attacks on form and procedure to undermine He's position. Zou declared that He should be arrested and returned to his exile. Local police officials, however, did not carry out Zou's orders, claiming it was not within their authority. Zou's shamelessly bold attitude must have come from a sense of his power on the local scene through his connections with Sun; it was, after all, the nonofficial Sun on whom Zou relied to make known He's supposed official misconduct.

Meanwhile, word reached Zou that He intended to send the real pardon that he had received with a memorial to the throne. He's inability to this point to use it in the county to prove the falsity of Zou's claims simply underscores the lack of credibility He had and, obversely, the power of Sun. Zou arranged a meeting with one of He Shunbin's disciples, an education official named Tong Xianzhang, who was home for a three-year period of mourning for his father.[36] Careers of Chinese officials were often marked by several-year interims to fulfill the proper mourning procedures. Zou apparently wanted Tong to help in devising some way to deal with the problems. Tong, however, well aware of Zou's duplicity, rebuffed him.

Zou's reaction perhaps more than any other event in this sordid story justifies the epithet "ruthless" as applied to the magistrate. In great anger at Tong but with great genius, he fabricated a story that Tong, on mourning leave, had been discovered digging up graves seeking treasure; such a sacrilegiously ghoulish act in a society that emphasized filiality and propriety was scandalous. Zou had Tong arrested and charged, and he then asked the judicial officer in Shaoxing, the prefectural seat, for death by strangulation. The judge had strong doubts about the charge, however, and dismissed Tong and the charges.

The brilliance of Zou's strategy depended not on the reactions of this prefectural judicial officer but in his assessment of how to mobilize the area's people to his side. While they might never be aroused by two government officials battling over esoteric issues that involved official seals and pardons, the image of someone scavenging graves, disturbing corpses, and stealing from ancestors was titillatingly horrifying. If this is the action of the disciple (Tong), then, one can hear them saying, what must the teacher (He) be like. The answer was clear: a stealer of official seals, a man arrogant enough to ignore proper official power, pretend a pardon, and return to his native place as one with authority.

Knowing that Tong would go to He's residence following his release, Zou dispatched county clerks and runners to He's house. They were joined by remnants of military units and area farmers, who approached He's house shouting: "Shunbin, you have rebelled against the emperor; we will imprison you." The chanting mob of perhaps several hundred people, brandishing agricultural tools and implements, tore down the door and entered the compound. They seized and bound both Tong and He. They stole He's official seal and the pardon that he had received. After looting his house, they grabbed both men and took them to court. Because we do not know the source or composition of the attacking crowd, it is difficult to gauge the dynamics of the situation. If these were area farmers without particular ties to Sun, enraged by the alleged evil of He and Tong, it becomes a case of people striking out at the man who was trying to salvage the source of their livelihood—a case of enthrallment to local powerholders at the expense of their own best economic interests. If, on the other hand, they were people with specific connections to the Sun family—tenants or workers in the brick and tile industry—then their actions were obviously self-interested. In either case, they did not see or understand the long-term economic necessity of conserving the lake and sought instead short-term security or satisfaction. The episode points to the power of the Sun family in lake affairs, the power of an unscrupulous magistrate who could manipulate opinion, and the galvanizing power of the idea of cannibalizing the dead. It suggests that political opponents have traditionally been bested in Chinese culture by the imputation of gross immorality.

Both Tong and He were flogged forty times. Tong was thrown into prison.[37] He Shunbin, however, met a much more tragic end. Ordered by the provincial court into exile in Qingyuan, Guangxi province, without appeal, He was fettered and led roughly away under the custody of an eleven-man escort led by one Ren Guan. In the court of the provincial judge, it turned out, were friends and confidants of Zou Lu; the administration of justice, like political and social affairs in general, depended in the main on personal liaisons rather than impersonal laws. After they had left, Zou Lu dispatched with special instructions thirteen men to catch up with He and his escort. The pretext was the fear that He's son, Jing, and daughter (married to a Fujian provincial official), who had not been arrested and were in hiding, might seek to rescue their father on the road. Driven by Zou's secret orders, the pursuers rapidly trailed the prisoner to the southwest. Entering Jiangxi province, they even jettisoned their supplies to be able to make better time. On July 19, 1498, they finally caught up with He and his captors as they spent the night in the Changguo Temple at Yugan; there they attacked and stripped He Shunbin, stuffed soaked clothing in his mouth, and suffocated him. Zou Lu hushed up

He's violent end at Changguo Temple when He's body was brought back for burial; the full story only emerged later.

The charge of Wei Ji to his protégé He lay unfulfilled amid the spasms of violence that shook Xiang Lake that summer and eventually snuffed out He's life. Certainly He had neither shrunk from controversy nor shirked Wei's instruction. But the challenge to the lake had changed from that which Wei faced three decades earlier: the power of lineage leaders and their allies had increased dramatically, and they seemed by the year 1500 so strong that one wonders whether even Wei could have handled them.

The story of the He-Zou controversy does not end here.[38] Zou had wanted the arrest of He Shunbin's wife and his son, Jing; but they fled to the home of Wang Ting in Changshu, Jiangsu province. Wang, a longtime friend of Shunbin, the two having received their civil service degree the same year, had served on the Board of Punishments in the central government. In obligation to his murdered friend, Wang took it upon himself to serve as a kind of teacher-father to Jing, inculcating in him the essentiality of avenging his father's death. Wang had taken Jing and his mother in; as Jing had a filial obligation to be the agent of retribution for his father's death, Jing also had an obligation to repay Wang for his friendship and guidance. On the wall of his room, written in blood, were the characters *baochou*—"Revenge!" Each day Jing would sit in the middle of the room, repeating endlessly the two syllables; every night Wang would call from outside his door, "He Jing must pay off this grudge," to which Jing would assent, firmly and with respect for Wang's training.

In late summer 1500, word came to Jiangsu that Zou Lu was being transferred to Shanxi province to serve as secretary in the offices of the provincial judge. Jing asked Wang if he could return to fulfill his obligation of revenge. Wang agreed and told him that he should seek help from his relatives. Wang himself offered to have some soldiers in his control follow and watch out for Jing. At a special farewell meal in the courtyard of Wang's residence, dice were thrown for a sign about the auspiciousness of Jing's departure. "Six-red" was thrown, and it seemed the gods had approved the undertaking.

When Jing arrived in Xiaoshan, it happened that the seal for Zou's new post had already been delivered to Hangzhou; Zou was to depart soon to receive his official credentials. Jing immediately went into hiding at the house of the lineage leader, He Ning. It was clear to both men that they must act quickly; first they needed the support of relatives. Without divulging Jing's presence, He Ning invited several tens of relatives and close friends for an evening of wine drinking. Whether the mood of the evening had been convivial is uncertain, but when about half of the available wine had been consumed, Ning began to weep quietly, saying, "Zou Lu's actions! Shunbin detested him." Continuing to weep, he called out, "What is to be done?" The bitter tragedy of Shunbin, an outstanding lineage leader, the display of Ning's deep emotion, and undoubtedly the effects of the wine brought tears to all at the gathering. He Ning raised his goblet to drink more, saying, "If Jing were here, we would have to avenge Shunbin; now, alas, where is he?" The group wept openly, many of them having become drunk.

At that point, the wine having helped to move the lineage members to a state of readiness, He Ning rose and said, "The matter is urgent: He Jing is here now! Are you willing to follow him in avenging his father's death?" They all answered, "Yes!" It is a strange scene: one wonders why, if He Shunbin had been such a respected member of the lineage, it was necessary for He Ning to get the group drunk before they would agree to act; likely the prospect of an act of revenge against an official of the emperor required some additional wine-assisted screwing up of courage. In any case, only after they agreed to act did He Jing emerge. He knelt before them, murmuring, "I am here; I am here." He followed Ning's lead, asking them if they would be willing to give their own lives to avenge Shunbin's death. Without hesitation, they agreed.

On the next day the group, dressed in white mourning clothes and armed with weapons, went to wait for Zou to pass on his way from the county seat to Hangzhou. The ambush site: Chenxi Garden. It was early autumn, most likely the period of irrigation from the lake. Probably still very warm, the weather on that day may or may not have reflected the violence that would occur, though the holistic Chinese world view would perhaps have been served had a storm brought destruction. The maples and catalpas would not have begun to change color, and the world of nature was predominantly arrayed in shades of green. In the garden, the greens may have been studded with red daylilies or the fragrant white banksia roses that would have matched the mourning garb of the attackers.

In their hands they carried staffs; Jing himself concealed an iron hammer and a sword in the sleeve of his robe. We are not told how long they waited for Zou and his escort to pass; but when he neared their hiding place in the garden, the ambush began with terrifying wails loud enough "to shake heaven and earth." Some in Zou's startled mounted escort fell prostrate on the ground; others fled. In his sedan chair Zou was stripped and received repeated blows from the group's staffs; he was so frightened, according to the sources, that his hair stood on end. Jing drew his sword during the melee. Intending to kill the magistrate now petrified from fear, Jing slashed his left thigh, but some of the group intervened, stopping Jing from killing him. The attack was over as quickly as it had begun; the garden was quiet. Zou, battered and bleeding, lay helpless.

Seizing him, the group crossed the Qiantang River for Hangzhou and a visit to the provincial judge in order for the lineage elder to state their grievance before this representative of the emperor. Even here, the group ran into problems: the official with whom they first had to deal, Xiao Zhong, was a close friend of Zou Lu. Enraged about the treatment of Zou, Xiao berated Jing and had him tortured. Jing was beside himself with anger (and perhaps pain); he screamed, "You must want to kill me, but I am not one who fears death. Don't you have a father and mother [whom you might have to avenge if necessary]? I already have demanded justice at the court. You can't slay me without the approval of higher authority." So wrought up was Jing that he savagely bit a chunk of flesh from his own arm and threw it on the judge's table; he then spat the blood from his mouth into Xiao's face. Everyone at the court was stunned by this action, which abruptly ended the audience. In any event, Xiao did not enjoy paramount authority in this case.

Jing, probably on the advice of his father's friend on the Board of Punishments, had already prepared his case in writing to the Board of Punishment's senior secretary Li, who at that time was concurrently serving as supervising censor. Li and the Shaoxing area circuit censor Deng Zhangza heard the case. They determined that Zou was guilty in his collusion with the lake encroachers and in the death of He; they sentenced him to death. But Li and Deng also judged the lineage attack on the former magistrate to be the symptom of a "dangerous illness"—disrespect for imperial officials. Jing was also therefore sentenced to death by strangulation. The several hundred people who were involved with Jing, it was determined, had different degrees of guilt. Jing's mother beat the drum outside the court as manifestation of her grievance against a system that had allowed her husband to be destroyed and would now destroy her son. Zou Lu also sent people to make accusations against the lineage.

Eventually the head of the Grand Court of Appeals, Cao Lian, was appointed to rehear the case with circuit censor Chen Quan. In his interrogation of Jing, Cao was very pointed: "Why did you beat up the magistrate?" Jing's reply echoes the substance of the famous Confucian analect on the preeminence of filiality over obligation to the state.[39] "I am not aware of a county magistrate; I am only aware of my father's enemy." Then, as if almost in apology, he continued, "I hate him; my only regret is that I still have not killed him." Cao was determined to get to the root of the revenge attack. He ordered that He Shunbin's body be exhumed and examined, since what had happened at Changguo Temple was as yet generally unknown and there was no proof of the cause of Shunbin's death. The medical examiner reported the wounds on He's body. At that time Ren Guan, who had been in charge of He's escort, came forward with the story of what had happened that night at the temple; moreover he produced a letter supposedly written by Shunbin in his dying moments. Ren is described in the sources as "public-spirited," but his silence about He's tragic death for over two years suggests otherwise. On the other hand, it also perhaps clearly emphasizes the tyranny that Zou Lu and his party held over the area and suggests the extent to which subordinates—whether government functionnaries or people in general—were cowed and intimidated by their political and social superiors. As a result of Cao's investigation, Zou Lu was sentenced to death by beheading and Jing was to be banished for three years.

There was, however, again considerable legal discussion about Jing's sentence, for he had attacked the emperor's servant. Many argued that it was Zou Lu who had become like a "dangerous illness," that Jing was only acting in retribution for his father and was therefore justified. At stake here was the confrontation between obligations: one to the state; the other to the family. At its heart was a dispute over the relative legitimacy of different types of authority among a people whose moral training stressed subordination to authority.

In Jing's case, the highest authority, the emperor, played an important role. The Hongzhi emperor (r. 1488–1506) or those around him decided Jing's crime was serious enough to necessitate as indefinite exile, which began in the early spring of 1501. But in the reign period (1506–1522) of the Zhengde emperor, both Jing and Zou Lu (whose death sentence had apparently not yet been carried out) were pardoned in one of the court's general amnesties. Jing returned home, where he died in 1514 and was buried next to his father. Many years later, the grave markers having long since been destroyed, those in the area were unable to remember which grave was father's and which was son's.[40] It seemed almost to be like the obliteration of the memory of the Pavilion for Viewing the River and the Lake or like nature's growth concealing the fact that there was ever a Chenxi Garden. Even before Jing's death the general amnesty from the few strokes of the emperor's vermilion brush had, for the purposes of law, equalized—and thereby wiped out—the moral differences between Jing and Zou.

And yet Jing became an enduring symbol of the filial son; his obligation to his father had been the obsessive and shaping force of his life. In a culture that emphasized obligation to father even to the point of death, the goal of his personal sacrifices (*not* those sacrifices themselves) elevated him to a position of great respect and honor. Jing, it is significant to note, lost his individuality in avenging his father; historical sources most often refer to him not as Jing but as "He, the filial son." The individual found his meaning only in his social position of son and in fulfilling that position flawlessly. After death, individual and social position were transmogrified to moral model. The creation of moral models to be inculcated by young and old alike was crucial in a civilization built on the subordination of young to old and female to male.

Among all the worthy defenders of the lake, Wei Ji and He Shunbin among them, it is Jing who is most clearly remembered in the annals on the lake, even though he had nothing to do with the lake itself. In the late nineteenth century poems were still being written extolling Jing and that "ancient drama of filiality."[41] Fu Tingyi, a scholar who wrote in the wake of one of the area's most devastating experiences, the Taiping rebellion, mused on violence and retribution.

> I visualize Chenxi Garden that same year:
> Relatives and friends all arrayed in white clothes and caps, pulling their hair, eyes weeping.
> [With such memories] people cease being startled that beheading the enemy is the method of retribution.[42]

Magistrate Yang Duo came to Xiaoshan in the aftermath of the tragedy and set out to deal with the Sun lineage encroachment on the lake, the issue that had precipitated the violent events. It would be a mammoth job: 1,327 mu (201 acres) had become paddy land; there were dikes with ninety-six openings and pools that divided the land into twenty-six strips; and there were 210 kiln households living on former lake land. Mao Qiling suggests that all this was restored to lake, that the prefect appointed a manager to help the county to return to the proper principles, and that eight elders were named to assist in the realization of these goals. But there is no other record that these reforms were realized. For the traditional crowning—and by then, it should have been seen, futile—touch, Yang ordered prohibitions carved in stone against encroachment, planting rice, draining water, and building on the dikes. The penalties ranged from exile to death. The irony was, of course, that less than three years earlier Shunbin had suffered both exile and death for trying to carry out Yang's very stone-inscribed objectives.

III

VIEW

The King of Yue Garrison, Late Sixteenth Century

CHAPTER

Strange Fruit: Lineages and the Lake, 1519–1555

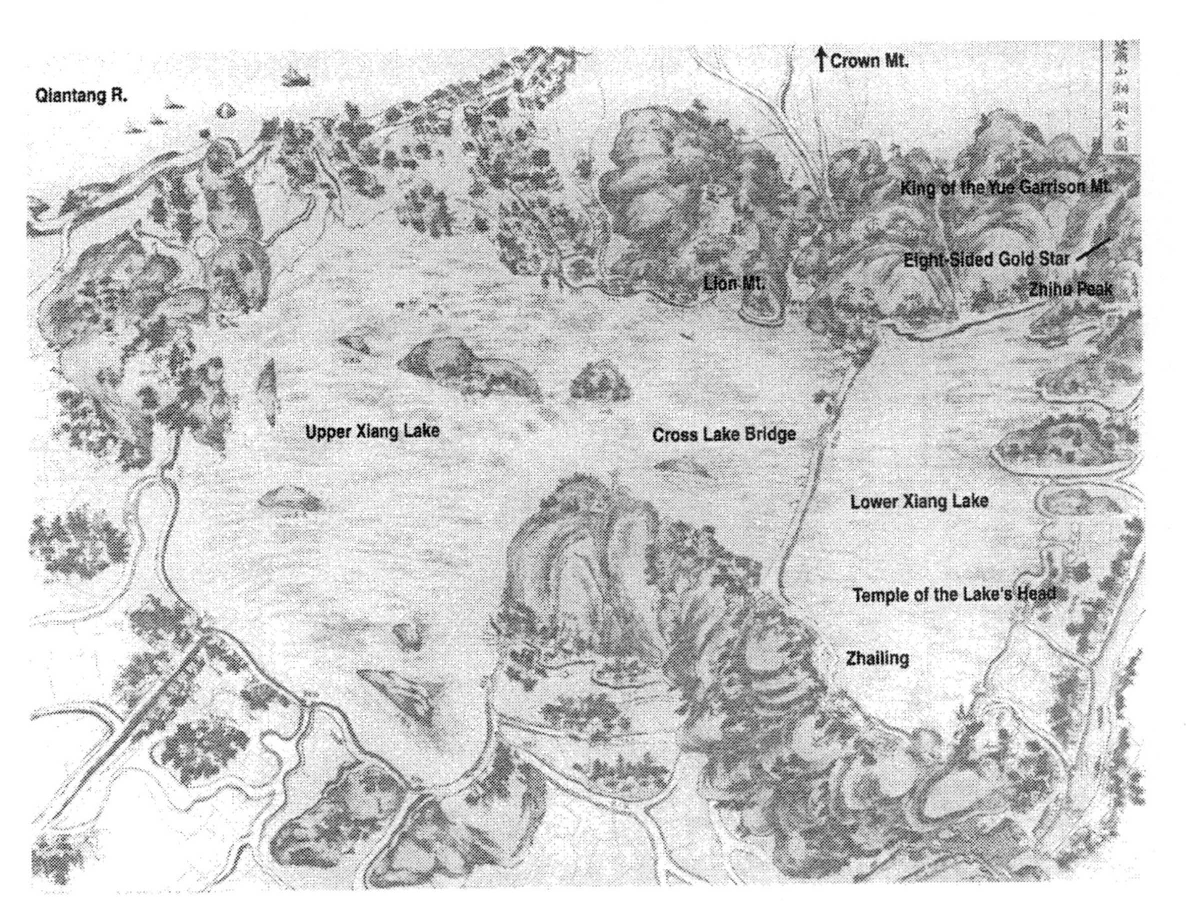

Map 7. Xiang Lake Sites, Mid-Sixteenth Century

VIEW: *The King of Yue Garrison: Late Sixteenth Century*

On the northwest shore of Xiang Lake, just over three miles southwest of the county seat, was a mountain known as the King of Yue Garrison, a symbol of military power (figure 2). In the late Spring and Autumn Period (722–484 B.C.), a lasting feud developed between the states of Wu, with its capital in present-day Jiangsu province, and Yue, headquartered in the Shaoxing area. In 496 B.C., the king of Wu was beaten by Kouqian, the king of Yue, and died as a result of wounds suffered in battle.[1] In the manner of He Jing much later in the lake's history, the dead king's son, Fuchai, set out to avenge his father's defeat and death; he succeeded in 493 by defeating Kouqian in battle.

Instead of listening to his faithful and farsighted minister, Wu Zixu, who counseled killing Kouqian and reducing Yue to a dependent province of Wu, Fuchai allowed Kouqian to live, gave him a small amount of land, and made Yue a vassal state. Wu Zixu's repeated remonstrance with the king only brought the king's irritation, consternation, and, in 485 B.C., order to commit suicide. Certain of Yue's eventual defeat of his lord, the dying Wu Zixu, the model faithful servant, is said to have cried, "Plant a catalpa on my tomb; when it can serve as the haft of a tool, tear out my eyes and place them on the eastern gate of Wu so that I can behold the annihilation of Wu by Yue."[2] Legend also has it that the tidal bore, which could easily be seen from the garrison, was linked to Wu Zixu; on his death he instructed his son to "take the skin of a ti fish, wrap it in my

Figure 2. King of Yue Garrison Mountain (1986)

corpse, and fling me into the river. Then, at dawn and dusk I shall come on the tide to gaze on the fall of Wu."[3] It is said that one might see in the white breakers of the tide moving up the bay with "wrathful sound" Wu Zixu in a death car pulled by white horses.[4] After determinedly fueling his resentment ("sleeping on firewood and tasting gall"), Kouqian defeated Fuchai in 473 B.C., thereby fulfilling Wu's prophecy. During the fighting Kouqian defended the mountaintop garrison that gave the mountain its name, and it became a symbol of his power and, some said, his tyranny.

Almost two millennia later, Magistrate Wu Shu, whose tenure preceded by two decades the He family tragedy, traveled to the King of Yue Garrison.[5]

> [From the foot of the mountain] I directed my sedan chair south past a small mountain stream. In the stream was a sight that struck me as unusual: there were several stones upright but leaning slightly in the water in the direction of the southward current. Now on foot, I entered a pine forest. Underneath the trees on a little path were a pair of cranes. When they saw us, however, they flew off with loud cries.

Though Wu's account is descriptive and not metaphorical, his narrative gives us an entree into the world of late Ming Xiang Lake and its denizens. The images of the cranes and their flight points us to the way Chinese scholar-poets in late imperial China dealt with the King of Yue Garrison. For the Chinese, the crane was a symbol of longevity; in Wu's account, they did not remain on the mountain but flew away. For later Chinese, the garrison became a symbol of the ephemerality of military strength and political power. Kouqian defeated Fuchai, indeed; but within a few years all that remained of the fierce battles and the cunning plans were ashes and ruins. Poets wrote of hearing ghostly sounds of horses arrayed for battle in areas around the mountain.[6] The obliteration wrought by time: the theme that typified so much of Chinese poetry. In a culture whose measure was human beings and their accomplishments, the pathos of human experience was painfully acute to the intellectually sensitive.

> Perhaps half a *li* in, there was a green stone cliff upon which climbed old vines. The path suddenly became dangerously steep so that in only a few steps of the climb, I could not straighten my legs and my feet hurt. So, I leaned against a stone to rest. When I felt somewhat revived, I got up and walked again. After continuing a long time, I heard [near the top of the mountain] the sound of flowing water. I proceeded from the ridge into a type of valley. Here was the Washing Horse Spring. The spring had formed a pond.

In the pond were produced fish and shrimp. When Kouqian opposed the king of Wu, the latter sent to him a salted fish.[7] The meaning attached to the gift was that Kouqian's state of Yue had many mountains but little water, that if Wu maintained its campaign and Yue did not yield, it would eventually die of drought. Even if the story is apocryphal, it points out the perennial problem of drought in the area. But the conclusion of the story was that Kouqian took a carp from the pond and sent it to Wu, evidence that Yue had water and could withstand a long siege; supposedly as a result, Wu lifted the siege and retreated. The symbolism of the two fish at the King of Yue Garrison—one salted, the other fresh—aptly pointed to the ecology of the Xiang Lake area; just as the pond provided the carp, so the lake provided life, preventing the barrenness of drought.

> Beyond the pond are two peaks which together look like earthen gates. After that the ground levels off; walking several more tens of steps, I arrived at the top of the mountain. It appeared that there are four tall walls—like military breastworks—with a lookout tower to warn about attack.

Though the garrison ruins existed from almost two thousand years before, their function still pointed to the great significance of defense for the people of Xiaoshan. The Qiantang River lacked the breastworks of a garrison, but it had often prevented enemies from entering the county—as in 1121 and 1352. But records indicated that threatening battles had occurred in or at the river in previous years as well: A.D. 196, 498, 590, and 882. As the editor of the 1693 county history put it: "We may not use the military for a century, but we can't forget it for a day."[8] For this reason, attention had been paid to the southern river bank defense and defense from the bay to the north.[9] If one looked from the garrison toward the county seat in the late sixteenth century, a new tall and thick city wall, indicative of renewed attention to defense, would have been starkly visible.

> [Also at the top] was a large spring, always full, commonly called Buddha's Eye Spring. Beside the spring is a small building of several rooms to worship the Buddha. An old monk of, say, sixty or seventy years (*sui*) was having a meal of boiled bamboo shoots.

Though the Buddhist establishment never reached the power it had under the Five Dynasties and the Song period, the Buddhist presence around the lake continued with the worship, festivals, and fairs of larger Buddhist temples and monasteries in easily accessible sites and smaller temples in mountain recesses and grottoes. Though in some quarters—especially among Confucian scholars—the social role and status of the Buddhist clergy was viewed with open misgiving, members of many families had connections with the Buddhist establishment.[10] From the late sixteenth century, two new, small Buddhist temples could be seen from the garrison: the Return to Prosperity Temple on Zhihu Peak, to the northeast, and, between it and the garrison, Refuge Peak Temple, located on a mountain known alternately (and poetically) as the Mount of Ten Thousand Silk Parasols or the Eight-Sided (*ba mian*) Gold Star.

> At the time the sun had already set, and I was also very tired. I rested on the couch [in the Buddhist hut] and fell asleep.
>
> At daybreak, I got up; and after putting on my clothing and cap went out for a view of the panorama. I could see the red glow of the rising sun reflecting in the sea.

Having seen the sunrise from the top of the mountain, Wu descended. Like the appeal of North Trunk Mountain (173 meters high) for those wanting a view from the heights, the King of Yue Garrison (169 meters high) offered similar delights. Early Qing dynasty scholar Cai Zhongguang noted from the summit the fascinating view of the sea, the lake, and the river with its dramatic turns.[11] Zhang Yuan, also of the early Qing, described the rain-shrouded view from the top in which the outlines of five small lakes to the west were blurred.[12] Mao Wanling, brother of lake chronicler Mao Qiling, used the epithet most commonly picked by poets for the heavens to describe Xiang Lake from the heights of the garrison: "kingfisher blue."[13] Writing in the early Qing, Mao also could have described, though he did not, a new bridge cutting across the lake; from the top of the mountain, it would have appeared a ribbon of brown running east-west and cutting the blue expanse in two.

In his account of his trip up the mountain, Wu records what he did immediately before his climb: "[There was the mountain of the] King of Yue Garrison. On top of the mountain was the garrison. At the foot of the mountain was a village of the Sun lineage. When I arrived, an aged lineage elder—more than ninety years old—welcomed me."

It is with a fine sense of historical appropriateness that at the foot of the mountain symbolizing power and tyranny lived the Sun lineage. The aged elder, though he might have had "shaggy eyebrows and hoary head" as Wu describes him, headed a lineage that was anything but enfeebled; it was indeed as determined and resourceful as Kouqian had been two millennia earlier.

CHAPTER: *Strange Fruit: Lineages and the Lake, 1519–1555*

The acute Chinese sense of the passage of time and history has not only meant obsession with the world of the past and its actors. Equally fascinating has been the unknown future, decipherable through the medium of portents, those peculiar phenomena—strangely colored clouds, earthquakes, comets, sightings of benevolent dragons—that were perceived as nature's way of foreshadowing events, either beneficial or catastrophic, that would occur in human civilization.[1] The unusual property of portents is that, although they foretell the future, their efficacy can be known only by looking at the past.

The historical record indicates that in 1551 plum trees in Xiaoshan county bore oranges. It was a bad portent: an agricultural handbook explained that when "trees produce strange fruit, there would in reality come a great calamity." More specifically, it warned: "When trees produce what is inappropriate, the farmers and all the people will be robbed." Or again, "When there are evil plans being hatched, trees cannot produce what is right." The Xiaoshan county history suggested that the oranges were a portent for increased taxation and for the local government clerks' using this increase as an excuse for robbing the people.[2] But in the long scheme of things, the portent more appropriately applied to Xiang Lake. By 1551 the lake had provided water to the area for over four centuries; while generations had come and gone, the lake endured. Then in the short period from 1553 to 1555, events occurred which many people believed help set the lake on its path to ultimate disaster.

The Three River Sluicegate

Almost two decades before the oranges appeared on plum trees, the lake community had been altered dramatically by the construction of a major water control facility for the drainage basin in which Xiang Lake lay. In the mid-fifteenth century, the Qiyan Mountain excavation, the building of the Maoshan sluicegate, and the diversion of the Puyang River to the northwest were momentous achievements. Most commentators on the water conservancy of the area, however, point to the Three River Sluicegate, constructed in the period 1536 to 1539, as the area's most important water control facility.[3] (See map 6, chapter 2.) Located where the old channel of the Puyang River entered the ocean, this sluicegate played a crucial role in the drainage of Xiaoshan north of the new Puyang River channel and much of Shanyin and western Guiji counties as well. This area was surrounded by dikes: from Maoshan to Xixing was the West River dike, crucial to prevent floods from the south and southwest; from Xixing to Songjialou, northeast of Shaoxing city, was the North Sea dike; and from Songjialou to Haoba, south of the town of Cao'e, was the East River dike. Although along the North Sea dike there had been two sluicegates for the drainage of floodwaters from the area within the dikes, they were insufficient to drain the area without an inordinate loss of life and property. Many times in the past (the first, recorded in 1181),[4] the dikes had to be broken in several places to allow the floodwaters inside the dikes to drain. The tragedy was compounded by the aftereffects of such a policy: ocean tide would flood in from the other direction, inundating once again the area near the dikes.

The large Three River Sluicegate, 102 meters long with twenty-eight drainage openings, was to regulate the retention of water within the dikes (figure 3). Constructed by prefect Tang Shao'en, the gates could be opened if there was too much water and closed during periods of little rainfall. As time progressed, other sluicegates were built on streams draining to the Three River Gate, and the area became dependent on the system for its very prosperity.

The finely tuned interdependence of the components of an operating efficient drainage system is remarkably fragile. In an area where streams were described as being as close and intermeshed as a spider's web, the condition of each stream's embankments, dikes, sluicegates, bridges, and the channel itself were intricately related. If weeds and silt clogged the stream, dikes alongside might not hold heavy rains; if a sluicegate leaked,

Figure 3. Three River Sluicegate (used until 1979)

irrigation water was scarce and the sluicegate's complete destruction in a flood was likely; bridges in disrepair might narrow the channel with collapsing debris; embankments used for hauling farm produce might be worn down and vulnerable to flood. Furthermore, each weblike stream was linked to the other web-streams to produce the Sanjiang system: too much or too little water from one, because of water's fluidity, would have immediate ramifications for the whole. The overriding objective was to maintain the system (dredge the stream; repair the sluicegate; reconstruct the bridge; clear the embankment) and regulate its deprivations and excesses through careful surveillance of sluicegate and drainage channel openings and closings.

While the mid-fifteenth-century Qiyan project had sliced most of two townships off into perpetual separation from the lake, the Three River Sluicegate project much more drastically constricted the Xiang Lake irrigation community.[5] With the sluicegate's construction, the Little West River, which wound its way from east of the town of Linpu and served as boundary between Xiaoshan and Shanyin counties on its way to the sea, became an excellent regulated source of irrigation for townships east of the lake. Laisu, Zhaoming, and Chonghua townships found that irrigation from the river was more convenient; the first two townships bordered the river, while the third was sandwiched between them and the lake (see map 3). None of the three used water from the lake after the mid-sixteenth century. Thus, four centuries after the reservoir's construction only three townships—Xinyi to the southeast and Changxing and Xiaxiao to the west and northwest—used Xiang Lake's water exclusively; parts of three others—Anyang, Xuxian, and Yuhua—still used some lake water. Of the original mu irrigated from the lake, less than half remained to receive the lake's water.

Such a drastic reduction of irrigated land from the lake's irrigation system had far-reaching implications for Xiang Lake. It eliminated from the irrigation community thousands of people who had been vitally interested in the integrity of the lake system. With their livelihood no longer dependent on the lake, their interest in lake dike maintenance, the orderly opening of drainage channels, or the fending off of reclamation ended. Drainage openings on the east of the lake—Tong Family Pond, Huang Family Stream, and Stone Grotto Outlet—were no longer officially used; therefore those charged with dike and watergate surveillance no longer watched, though pilferage of water by people living near the lake increased. The system began to fail as certainly as if a leaking sluicegate on a small stream in the Sanjiang system had not been repaired. From that time on, had a resident at the foot of Golden Chaff Mountain, let us say, decided to haul baskets full of soil into the lake to extend his paddy, there would probably have been no immediate outcry from the irrigators of the three townships. From that time on, had a farmer of Stone Grotto Village damaged the dike by pulling a heavy load of decomposing water weeds and grass to fertilize his fields, there would likely have been no immediate protests. From that time on, had a wealthy man sought to take for private use lake land meant for the public, there would likely have been those to agree that this was of no consequence, that the lake could justifiably be smaller since there were fewer mu to irrigate. The three townships' river irrigation could, then, become a positive spur to lake reclamation. Not only did the community of irrigators shrink, but this reduction also gave rise to lapses and collapses in the system.[6]

The Xiang Lake Social System

In many ways the ideal of interdependence in Chinese society was akin to the finely tuned interdependence of the components in a drainage system. Just as a sluicegate had to be placed *in* a dike *across* a stream—that is, in relationship to other components—to be meaningful or useful, so too did an individual find his meaning only in relationship to others in society. Many kinds of connective links became significant in the formation of social networks and groups in Chinese society. The strength of the group or network depended not only on its function but also on whether it and its goals were buttressed by widely held social and cultural values, values that shared the overriding objective of a drainage system: maintaining the system.

Moving from the looser to the more cohesive, we may first point out the functional network of the Xiang Lake community of irrigators. Joined by a common goal—to attain enough water to grow their crops—farmers had a stake in the regulated and maintained state of the lake. But it was a *public* utility created for *private* ends. Irrigator Wang in Xuxian township had no linkage or obligation to irrigator Han in Xinyi township other than following the irrigation regulations; by doing so—and this was the concern of each—they could gain the needed water. But if, in any given year, a drought drastically lowered water in the lake, then Wang and Han, with no personal obligation to each other and desperate for water, might have few qualms about pilfering water or bribing the dike chief to leave the sluicegate open longer than regulations stipulated. The personal relationships in the irrigation network were not reinforced by the values inculcated through education, rituals, or state-sanctioned models. In such a large functional network, linkages were thus fragile and could easily be severed by natural exigency.[7]

Certain important social linkages could be extended into wider networks and brought a more general sense of obligation to bolster their strength. Receiving civil service degrees in the same year brought a lifelong linkage. The shared sense of achieving from the state the status necessary for particular service to the state and (through infusion of Confucian ideals) to the people fostered this "alumni" sense of commonality. Obligation came from personal connection; thus Wang Ting, a same-year degree recipient with He Shunbin, felt a deep obligation both to receive He Jing and his mother after the censor's murder and further to encourage the act of revenge.

Three similar social linkages important in the history of the lake were strengthened by reinforcement from deeply inculcated familial ideals. The link between people from the same native place brought mutual obligation and often promoted potent political alliances. In the Chinese context, the strength of this tie lay in sharing the same birthplace where respective ancestors were buried. The teacher-student or teacher-disciple relationship also had about it a quasifamilial aspect (that is, the intellectual father) that bolstered an already special relationship, given the Chinese emphasis on education. In the Ming period, the relationships of He Shunbin to Wei Ji and of Tong Xianzhang to He Shunbin clearly evinced the practical strength of this linkage. In the same vein, the relationship one felt to the dead was significant not only with biological ancestors but with what might be called functional ones: the relationship and obligation of people like Zhang Mou, Wei Ji, and He Shunbin to the Four Officials of the Song were a tangible force shaping their own actions.

The key sociocultural unit in the interdependent Xiang Lake system and therefore in the history of the lake from the early Ming to the twentieth century was the family, more specifically the family writ large in the lineage, a group that traced its ancestry to a common patriarch. Lineages in Xiaoshan, Shanyin, and Guiji counties were famous for their cohesion, power, and influence. Many villages were lineage villages, where all the inhabitants shared a single surname. Even larger towns were dominated by single lineages: most of the residents of Long River, two miles west of the lake, were surnamed Lai, while in the Puyang River's New Dike Town, less than three miles southeast of the lake, most were named Ni.[8] Lineages employed various strategies to build and strengthen their material power, but the crucial bolsterer of ties in a lineage were the cultural familial values inculcated from childhood.[9] The filial son saw after all the needs of the parents, fulfilled the proper ancestral rites, and produced offspring to carry on the ancestral line. Mutual obligations of father and son, husband and wife, and elder and younger brother were fostered by the essential familial bonds. The revenge attack on Zou Lu by He Jing and his lineage associates show the importance of these values in practical situations. A strong lineage maintained an ancestral hall; it printed a genealogy that set forth the line of common descent, providing tangible evidence of the unity of the lineage; and, in many cases, it established a lineage estate (*yizhuang*), the profits of which furthered lineage interests, funded annual rites, and paid for training young lineage males for the civil service examination.[10] In sum, of any social network or grouping beyond the family itself, the lineage had the material and sociocultural cohesion and strength to sponsor effective, long-term historical actors.

The Sun and Wu Lineages, Early Sixteenth Century

The Sun and Wu lineages began to rise to local prominence in the turmoil of the end of the Yuan dynasty. Though a man surnamed Wu had been instrumental in the lake's beginning in the Song period, we cannot tell if he was an ancestor of this lake lineage. The earliest recorded effort of leaders of the two lineages was the usurpation of lake for use as paddy land, a deed indicating their agricultural interests; that first encroachment was cleared in the Yong'le period (1403–1424). Wei Ji faced the lineage nexus in the mid-fifteenth century, reining in the Wus but leaving the Suns for He Shunbin to quell. The encroachment that initiated the Censor He affair was intended to fill in the lake area near the shore for further kiln construction in the expansion of the lake's most important industry, brick- and tile-making.[11] Made from Xiang Lake clay dredged by lineage members, the bricks and tiles were fired in kilns built by the lineages in villages along the lake and were sold as construction materials primarily in Hangzhou and Shaoxing.

By the last years of the Zhengde period (1506–1522), it had been almost sixty years since Wei Ji's dike renovation. Their state of disrepair not only made it likely that water would leak out or flood over the dikes during spring rains, thereby lessening the supply for late summer irrigation, but also made the digging of clandestine drainage holes and subsequent pilferage of water easier. The prohibitory characters carved in stone by Magistrate Yang Duo were dead letters.

Proper timing for encroachment was of the essence. Petty encroachers dug illegal drainage holes in the dikes at night. For more widespread intrusions into the lake, choosing a time when official attention was diverted to other problems allowed encroachers to present those around the lake with various faits accomplis. In the late Zhengde period, Sun Zhaowu and leaders of the Wu lineage took advantage of threatened political unrest in Hangzhou to act. A rebellion of Prince Ning (Zhu Zaihou) in Jiangxi province in July 1519 prompted the eunuch Bi Zhen in the Zhejiang provincial administration to attempt a coup. The coup was halted through the provincial surveillance commissioner's watchfulness, but the counties around Hangzhou remained on defensive alert during the summer and fall.[12] The Suns and Wus moved quickly, usurping lake dikes to plant crops, pasture animals, and even construct buildings.

Though there is no record that these encroachments were effectively rolled back, they did stimulate the announcement of new dike management regulations, which proved a notable change from Zhao Shanqi's reforms of 1158. Until then, officials had frequently met with area elites to discuss crucial issues facing the lake. After 1158, however, sources describe lake policy as devised and managed by county agricultural officials.[13] The absence of local elites from accounts of lake maintenance and regulation does not necessarily mean that they did not participate or were not consulted by officials. It does suggest, however, that their active participation in lake administration was probably subordinated to that of local officials. This period of general official lake management stretched from the late twelfth to the early sixteenth century. It likely signified an official distrust of the power of wealthy elites, but it did not, of course, always signify good management: Intendant Zhang of the Song and Magistrate Zou of the Ming are cases in point.

To reduce the power of local officials, which had swelled dangerously in the turmoil at the end of the Yuan period, the founder of the Ming dynasty had granted the officials' former tax collecting and judicial power to new local rural organizations, or *lijia*.[14] Given considerable power, lijia headmen held their positions for limited tenures, a check that lessened potential self-aggrandizement. The Xiang Lake regulations of 1520 set up a system modeled on the lijia principle.[15] The lake dikes were to be divided into nine sections, one for each township to control. Each township would select households to police the lake dikes, preventing encroachments that threatened the lake. The regulating households would arrange for two able-bodied men to become lake administrators (*huzhang*) and be sent to the designated lake shore. If dikes needed repair, a lake administrator would contact the people of his township. Every month the administrator would issue a report to county officials, who were to investigate any noted problems.[16] If county officials failed in their duties, it was the administrators' responsibility to pursue cases of encroachment through investigation and the arrest of the guilty. If administrators conspired with or allowed themselves to be bribed by wrongdoers, they would suffer the same fate—exile to the Manchurian frontier—as the lake encroachers.

Several features of the new regulations elucidate past and current lake problems. The township, not officials, bore the duty for routine dike repair and maintenance; this shifting of financial responsibility certainly reflected the general deterioration of the Ming fiscal apparatus in the early sixteenth century.[17] That administrators were to fulfill officials' functions points to the sorry record of past officials and to a major shift in the management of local affairs. Last, the lake administrators received relatively little reward for their time-consuming, onerous tasks. In ordinary times that would likely have soon made the regulations dead letters.[18] As it was, the local power situation under the Sun-Wu lineage nexus precluded effective reform. If the system worked at all we do not know: the sources are silent.

The Rise of the Lai and Huang Lineages

As the Sun-Wu power grew in the fifteenth century, other lineages near the lake also began to extend their prestige and enhance their wealth. The Lai lineage had lived to the west of Xiang Lake since the twelfth century, perhaps immigrating to Xiaoshan in the wave of refugees of the 1120s. Its villages had first been located at the base of Crown Mountain on the shore of White Horse Lake.[19] As lineage prosperity developed, the town of Long River sprang up, often less than a mile away from the Qiantang, but, with the river's unpredictable deposit of silt along the bank, sometimes much farther. The site was propitious for the lineage's fortunes. With sufficient water, the land around Crown Mountain was the best in the whole county;[20] established after the construction of Xiang Lake, the lineage base seemed to have solved its water supply problem. In addition to its agricultural potential, the town was less than two miles away from the flourishing river port of Xixing, whose main market was on the south bank of the Grand Canal and through which goods passed from Ningbo and Shaoxing in transit to Hangzhou.[21] The town was also only three miles from Wen Family Dike, an increasingly important entrepôt on the Qiantang. Though evidence is lacking that the Lais were immediate commercial successes, by the Ming period the lineage had numerous links to Xixing and Wen Family Dike businesses even as they continued to concentrate on agriculture. That necessarily meant careful attention to the river and sea dikes so close to their farmland; many of the essays and poetry of Lai literati focus on the river, especially its ominous sound and menacing power, and the dikes that had to contain it.[22] The famous philosopher Wang Yangming (1472–1529), after climbing nearby Lion Mountain, west of Xiang Lake, remarked on the sound of the swiftly flowing river as it moved to the "ocean's gate."[23]

By the mid-fifteenth century, members of the Lai lineage had sufficient wealth and stature to be included in the county history's biographical section. Lai Can was active in building dikes from irrigation from small lakes west of Xiang Lake, helping the area prosper. He also became the local contact for the prefectural administration in its reconstruction of area sluicegates, contributing his own funds and personal direction when official funding ran dry.[24] Lai Heng contributed money to repair the sea dike.[25] Lai Li contributed large amounts of grain to the area when asked by the government at time of famine.[26] Although the Lai lineage focused on maintaining river and sea dikes, they depended on Xiang Lake for the bulk of their irrigation, and the lake, as for all in the area, remained in its beauty a source of emotional and psychological pleasure. Lai Li's poetry about Xiang Lake intimates through its images the pleasant contentment of an old man having lived a satisfying life: his small boat floating on the lake, the haze of the late afternoon sun toward the river, the scattered white clouds, the cries of the crows—images of "good things, erasing [any] anxious thoughts."[27] On his death in 1459 no less a local notable than Wei Ji praised his vigor and public spiritedness.

Wei Ji also wrote the eulogy for Huang Xunli, an immensely wealthy merchant who died in 1456 after a distinguished career in charity and famine relief. Buried on Lion Mountain, which Wang Yangming had climbed to view the area, Huang hailed from Daishang, southeast of Xiang Lake. Like the Lai lineage, the Huangs experienced in the fifteenth and sixteenth centuries increasing wealth and reputation. Moving from Zhuji county south of Xiaoshan during the mid-twelfth century, the Huangs put together agricultural success (Huang Shoutang, who died in 1461, was known as the Ten Thousand Picul Elder) and commercial know-how to build substantial lineage strength.[28] Shoutang's son, Yangxing, spent much time and money in water control efforts, including repairing the West River dike and preserving Xiang Lake. Yangxing's biography also suggests that another lake lineage, the Han, to the southeast of the lake, had fallen on hard times. During the late thirteenth century the Han had been large landholders in the area; they had, it should be remembered, built the Pavilion for Viewing River and Lake and commissioned its formal renaming. But two centuries later, at least one of the Han was in serious debt to Huang Yangxing, unable to make his interest payments to Huang after bad crop years. The Huang genealogy records that Yangxing, in his benevolence, forgave the debt.[29] In spite of general increasing prosperity in the late fifteenth and early sixteenth centuries that could help raise the fortunes of lineages like the Sun, Wu, Lai, and Huang, not all lineages in the area necessarily shared the happy economic situation. For whatever mistakes in judgment, streaks of ill fortune, extravagant misuse of earlier wealth, or lack of effectiveness in business dealings with leaders of other lineages, the Han would have to wait another day to rise.

In addition to reliance on agricultural and commercial sagacity and expertise, at least two other strategies were instrumental in promoting lineage strength. One method was intermarriage, whereby the fortunes of two lineages might become more closely linked.[30] Leaders of the Sun and Wu lineages used this strategy effectively in maintaining and expanding joint control over the tile-making industry. In the fifteenth and sixteenth centuries the Lai also used marriage ties to the Wu and the Huang lineages, as well as to two other noted county lineages, the Wang and Cai, to bolster their local status.

A second strategy was attaining civil service degrees and the official position that could come with upper-level degrees (*jinshi, juren, gongsheng*); with position came automatic great prestige and opportunity for substantial perquisites and financial benefits that could trickle down or be funneled into lineage coffers. Developing since at least the Tang, the civil service examination was the vehicle to degree and position. It was pivotal in the decline of the centuries-old aristocracy of medieval China and the rise of the civil bureaucracy in the Song period and after. In the years after 1313, the commentaries of the famous Song philosopher Zhu Xi had been accepted as the orthodox, "official" interpretation of the classic texts that young males should master. All officials in the realm were thus inculcated with the social and political values embedded in Zhu Xi's thought. Joined with the sociocultural values of familiality, the ideals of social harmony and service to the people became hallmarks of Chinese social and political thought.

Not all lineages emphasized every strategy in their approach to social position or executed every strategy successfully. The traditional view, that the civil service degree became the sine qua non for social power and mobility, may have been true for families with pretensions to power beyond the locality, but Xiang Lake lineage histories record a different story for local affairs. Some lake lineages, the Lai, for example, had an outstanding success story in degree attainment.[31] In the fifteenth century the Lai lineage produced seven upper-level degree holders, and in the period from 1525 to 1620, no fewer than thirty-two.[32] In addition, from the 1430s to the end of the Ming, they produced twenty-six purchased degree holders (*jiansheng*) who had opportunity for official position. The Sun lineage, in contrast, while producing four upper-level degree holders in the 1400s, only produced five in the sixteenth century, and the Wu lineage only had two for the entire sixteenth century.[33] And yet the Sun and Wu lineages in the fifteenth, sixteenth, and seventeenth centuries had the local power to intimidate all local resident elites as well as local official elites. At least in the Ming and early Qing dynasties, the civil service degree was not necessarily the essential component in the equation of local power. It is probably safe to say that all lineages aspired to at least some examination successes (even the Huangs, who, it was said in the fifteenth century, lacked the intellectual ability to succeed in the examinations).[34] But it is clear that other strategies could be equally or even more important.

Attack from the Japanese Pirates

Into the world of lineages and their strategies for enhancing financial status and prestige, into the yearly cycles of planting, weeding, irrigating, and harvesting, into the continual concerns about the unpredictabilities of the water supply, came, as unexpectedly as oranges on plum trees, a series of brutal raids from the sea, invasions led by so-called Japanese pirates. The problem had indeed begun with swift attacks by Japanese pirates who looted communities along the coast with impunity and then retreated.[35] The dilapidated state of the Ming defense only invited more attacks, and the booty lured some coastal and inland Chinese to participate in raids that intensified in size and number. By the early years of the 1500s, most pirates were Chinese.[36] While first attempting to deal with the problem using local forces, the Ming, seeing the raids take on the proportions of repeated invasions near midcentury, eventually appointed a special supreme commander to oversee the defense of southeastern China.[37]

From 1553 to 1555, Xiaoshan was attacked from three directions by marauding bands. The first attack, in late May or early June 1553, was largely repulsed on Xiaoshan's coast at Turtle's Son Gate, where the Qiantang entered the sea between Kanshan and Zheshan, several miles northeast of the county seat; but some pirates penetrated into Xixing and killed a number of people.[38] One of the county's first defensive reactions was to construct a wall around the county seat; about twenty-nine feet tall and twenty-four feet thick, it promised to deter would-be attackers. Begun in December 1553, the wall was completed by April 1554. Meanwhile, a military training ground for existing local forces was established on the shore of Xiang Lake.[39] For those writers who often reflected on the military machinations and paraphernalia of Kouqian associated with the King of Yue Garrison, the existence of military training at the lake must have had the eerie quality of a déjà vu. The training ground, though constrictive, was used until 1556, when it was moved to wasteland north of the county seat.

In late September and October 1554, many pirates invaded the Ning-Shao area, including Xiaoshan, looting and burning houses, indiscriminately and brutally killing inhabitants, and destroying bridges and public property. Their primary targets were the elite dwellings that dotted the plain. Surrounded by high walls, these quadrangular, tile-roofed houses were centers of comparative wealth in an expanse of general poverty. Residences of civil service degree holders had tall flagpoles before their entrance gates—marking them as lucrative booty sites for the pirates.[40] With the ineffective military defense of the Ming, the suffering population seemed, as often before, at the hands of forces they did not understand and could not control. Local defense often fell to the family or lineage itself. In the autumn attack, two Lai brothers, Duanmeng and Duancao, volunteered for the lineage to serve as scouts to track a group of pirates who had left Xixing and were following the Qiantang upstream. When the sixty-plus pirates came ashore for a raid, they were met by over one hundred young men of the Lai lineage who had hoisted a large silk flag on the road with the characters for "Lai lineage." The pirates, according to the sources, were so intimidated by the unexpected show of potential resistance that they proceeded upriver to other counties.[41] Whether the banner and Lai men were the compelling reason for their departure is impossible to say.

Unfortunately for the area, the Lais did not intimidate many pirates for long. In late November and December 1554, pirates attacked from the southeast after giving up their plans to seize Shaoxing. They were met by a large Ming force that captured and executed more than two hundred.[42] In June and July 1555, pirates were defeated in three separate Xiaoshan battles north of the Grand Canal. In November pirates holed up in Ding Village, several miles southeast of Xiang Lake, where a successful Ming attack led to the seizure and execution of twenty-six men. The final pirate battle in the county occurred shortly afterward when pirates, on the run, took sanctuary in a small stockade at the foot of Kanshan on the coast of Hangzhou Bay. Ming forces, who greatly outnumbered the pirates, attacked. In a bloody fight, the desperate pirates showered the besiegers with roof tiles, spears, and knives but were finally defeated.

From 1553 to 1555 fighting had ranged over the county, from Xixing in the west, to Ding Village in the south, to Kanshan in the north. The destruction was considerable; if judged by records of necessary reconstruction of public works, the majority of the damage was around the county seat and to the west of Xiang Lake.[43] But the psychological damage to the area's inhabitants must have been considerable: three years of a siegelike atmosphere with threats of brutal attacks looming continually must have taken a toll. When the Lai took decisive defensive action, it was certainly with the knowledge that the plunderers' target was generally the elite households where booty was available; but the populace as a whole suffered from the burning, raping, looting, and cold-blooded slaughter. One could surely take the sign of the oranges as a portent of the tragedy.

Xiang Lake Divided

But for Xiang Lake and ultimately its dependent population, a more serious development occurred simultaneously with the pirate invasions. In 1554, while the lake community and officials in the county seat focused on defense against the invaders, Sun Xuesi forever changed the shape of Xiang Lake by building a bridge across it at its narrowest point, effectively cutting it in two.[44] Sun, a former senior secretary on the Board of Rites and censor on the Grand Court of Appeal, took a page from the encroachment strategy of the previous generation's Sun Zhaowu and made his bold move when the community was least able or unable to react. The Cross Lake Bridge joined the Sun lineage home, Sun Village on the Lake on the lake's western shore, to Zhailing, the Wu lineage home on the lake's eastern shore (figure 4). The goal: easing the trafficking between the two lineages in their control and management of the brick and tile industry. The bridge also enabled Sun to reach the county seat more quickly. There is no question that the bridge saved time and brought the Sun and Wu lineages greater convenience. At stake was the use of a public resource for private gain.

There seems little doubt that a period of rising economic prosperity in the late fifteenth and early sixteenth centuries in this area had brought handsome profits to the Suns and Wus in their manufacture of construction tiles and bricks.[45] The Suns had moved rather diffidently into the civil service degree strategy, but in the first half of the century they secured one jinshi and two juren degrees. Sun Xuesi reached official position through recommendation for his learning.[46] His brother, Sun Xuegu, earned the jinshi degree. Even the written history of the lake, the editors of which were no admirers of the Suns, calls them the lake's "preeminently able" lineage.[47]

Figure 4. Cross Lake Bridge (1920s)

To display his power and wealth, Sun Xuesi undertook many construction projects. In addition to the bridge, he directed the building of several Buddhist temples and shrines around the lake. Near the site of Chenxi Garden he built the Temple of the Lake's Head; together with two smaller temples, the Return to Prosperity on Zhihu Peak and Refuge Peak on the Eight-sided Gold Star, it symbolized his ties to Buddhism and represented the charitable display of riches and dedication certain to strengthen *individual* karmic wealth, if not that of the lineage.[48] In addition, as a further exhibition of material wealth and psychological self-congratulation, Sun and his brother built outside the west gate of the county seat an honorary arch called Twin Phoenix of the Lake Arch, an unsubtle comparison of the status of the two Suns to the phoenix, preeminent among birds.[49] The placement of the arch so near the seat of county political power and actually some distance from the lake for which it was named symbolized the Suns' political power over officials in the county and the lake area.

Whatever the Suns' record as builders, the lake bridge was destructive. In effect a dike with one rounded opening, it blocked water flow in the lake. Whenever irrigation water drained out of the narrow northern section, now called Lower Xiang Lake, a blockage of the equalizing flow of water from the southern section, Upper Xiang Lake, occurred. As time went on and water weeds and grasses grew along the bridge base, Lower Xiang Lake became shallow and itself clogged with water weeds; not all the remaining irrigators could then get the water to which they were entitled.[50] The bridge, in short, began to destroy the hydrological system of Xiang Lake.

Huang Jiugao, a scion of the Huang lineage southeast of the lake, jinshi of 1538, and retired secretary of the Board of Works, made a name for himself on the local scene through his study and repair of the West River Dike.[51] But he was aware of the bridge's devastating effect on Xiang Lake's future. A poem he wrote about the lake is marked by caustic double entendres.[52] He paints a picture of the lake as a water world of purity, depth, and fragrance, productive of water plants and fish, created to prevent drought. In an image of engendering life, Huang talks of streams of clear water rolling from the lake to slake the "earth cracked by drought." He describes walking in straw sandals around the shore of the kingfisher blue water on a beautiful autumn day. He mentions that looking in the direction of the Eight-Sided Gold Star, one could see the irrigators: Sun Xuesi was buried at the foot of this mountain and the juxtaposition is surely intentional. The phrase meaning "earth cracked by drought" is probably a double entendre for "[the work of Yang] the Tortoise broken" (*Gui-che*). One of the nicknames of Yang Shi, the creator of the lake, was Tortoise Mountain.[53] Huang concludes his poem,

> . . . then we come to the shrine for Yang;
> Directly opposite is Wei Ji's grave.
> Who has already narrated in verse the eight griefs?

Griefs could obviously refer to the deaths of Yang and Wei, two of the lake's foremost upholders, but the coupling of griefs with "eight" in the context of protectors of the lake seems to bring us back to the griefs caused by the destroyer, Sun Xuesi.

Huang's veiled critique of Sun's lake activities was not echoed by many contemporaries. People in the area, retired scholars and officials alike, with clear remembrance of the fate of Censor He Shunbin who had not been silent about the Suns, said nothing. Many actually praised the building of the bridge, calling it a natural division, since the lower lake drained north and the upper lake drained south. The Suns held such power that, despite the damage to irrigation, the area's people were cowed into silence or even intimidated into devising flimsy justifications to avoid angering the Suns.[54]

The Suns' degree of intimidation of the area's populace is most flagrant perhaps in their efforts to expunge from the historical record the tragedy of He Shunbin and the filiality of He Jing. The placement of Sun Xuesi's Temple at the Lake's Head at the site of Chenxi Garden was probably a symbolic attempt to obscure and overcome the symbol of the garden as the site of filial revenge that had been prompted by the machinations of Zou Lu and the Sun family. Sun transformed it instead to the site of a new, imposing temple.

An even clearer attempt to obliterate the past was the Suns' effort to delete the He family members from the historical record. Specifically, their goal was to revise the county history to exclude the Hes from the biographical section, to eradicate mention of He Jing's filiality, and to mention the He lineage actions regarding Xiang Lake only vaguely: "In 1499, a county man [He] memorialized. . . ." In 1537, as an expression of his own filiality, a grandson of Censor He, He Shifu, wrote an essay requesting those in charge of the revision to correct omissions, especially to include a biography of Censor He and to stress his father's filiality.[55] He claimed that "the death of my grandfather was the death of the lake; my father's demand for justice was for justice for the lake: what he wanted to restore was the lake." He Shifu's plea beautifully linked the twin virtues of public service and private morality, qualities not notably apparent in the Suns' own historical record. Those in charge of the new history held public discussions with elders from Yuhua township, the home of both the Sun and He lineages. Finally, there was agreement to include He in the biographical section and to restore his name in an outline account of the Xiang Lake affair (without naming any of He's opponents). He Shifu seemed to have rescued his grandfather and father from historical oblivion. But it was only temporary: the 1935 county history notes that for almost two hundred years—from after 1573 to 1750—no subsequent drafts of the county history included a biography of Censor He.[56] More than that, perusal of the 1693 edition, for example, shows that in the chapter on water conservancy the editors, though mentioning in outline form the activities of He and his son, do not name the Hes' opponents, and in their treatment of the 1519 encroachment efforts they fail to mention that the Suns were involved.[57] The Suns are named in official documents printed in smaller type and following the text, but clearly the Sun lineage was able to manipulate the main historical text for their own benefit.

Mao Qiling's account of the relationship of the Sun and He lineages reveals that the key event leading to the exclusion of He Shunbin and He Jing was the attainment of the jinshi degree by Sun Xuegu, Xuesi's brother. The sociopolitical legitimation of this highest degree in 1544 brought even greater clout to the Suns in their dealings with scholars who shaped the past. Mao indicates that the editor of the late-sixteenth-century county history excluded the He names in part because of He Jing's attack on an official but mostly because of the Suns' power and their intimidation of all opponents.[58] The Hes reappeared in the local history only with the general demise of Sun power: until then, the Suns were able to control not only the area but even the history of the area.

Sun's Legacy

Not much is recorded of Sun Xuesi's relationship with his brother, Xuegu. We know they joined in building the Twin Phoenix of the Lake Arch, but they must have been very different. Sun Xuegu, with jinshi degree, held like Xuesi, official positions, at one time serving as magistrate of Dongguan county in Guangdong province. In an irony that is so perfect it almost strains credulity, during Xuegu's magistracy a wealthy man built a dike in the opening (of a river?) to the sea. The dike was simply for his own advantage, and it was clearly injurious to the county's inhabitants. Xuegu, with all the energy at his disposal, had the dike destroyed. In gratitude, the people of Dongguan built a shrine for Sun.[59] Whether the two brothers ever discussed this episode, we are not told.

In addition to his construction efforts, Sun Xuesi left one poem: the theme, the obliteration of the past with the passage of time; the topic, the King of Yue Garrison Mountain:[60]

> Yue Peak, remote, from of old had a fortress
> With the King of Yue already garrisoned there, a military camp.
> In those years a tyrannical spirit [the state of Wu] put all to an end;
> But another day the beacon fires at Kanshan burned again, even brighter;
> Under enemy insults, the [King of Yue] nursed vengeance
> Until he could restore the fame of the garrison which lasts to the present.
>
> The moon is black over a thousand peaks.
> At night it is as if there were rank upon rank of soldiers engaged in ghostly battles.
> The sound of horses . . .

This musing on fame, successful revenge, and the echoes of power reveals a remarkable resonance with the themes of his own lineage affairs over the preceding half century. One wonders whether Sun sensed, in light of historical comparisons, the fleeting reality of his power. Certainly his construction of temples, shrines, and arches was one way to preserve his fame and that of the lineage; and by obliterating the actual past in the local record, he could obscure the unsightly and bloody blemishes of the past. In his lifetime and for six or seven generations after, Sun power remained undisturbed. But the black moon of obliteration lay inevitably ahead.

Sun died and was buried at the foot of the Eight-Sided Gold Star. Years later—we do not know exactly when—in an act shocking in its resonance of fragments of the history of the Suns and the lake (the grave-digging accusation against Tong Xianzhang, the filiality of He Jing and He Shifu, and the Suns' own efforts to delete the past), the descendants of Sun Xuesi broke apart and destroyed the ceremonial stone figures and decorations before his grave.[61] Through this surely unfilial desecration, his posterity, like He Jing, sought to avenge the shame wrought upon their lineage by this man. While they could destroy the tangible gravesite structures, his legacy, the bridge and its baleful effects, remained.

IV

VIEW

On Pure Land Mountain, 1689

CHAPTER

Image and Remembrance, 1644–1705

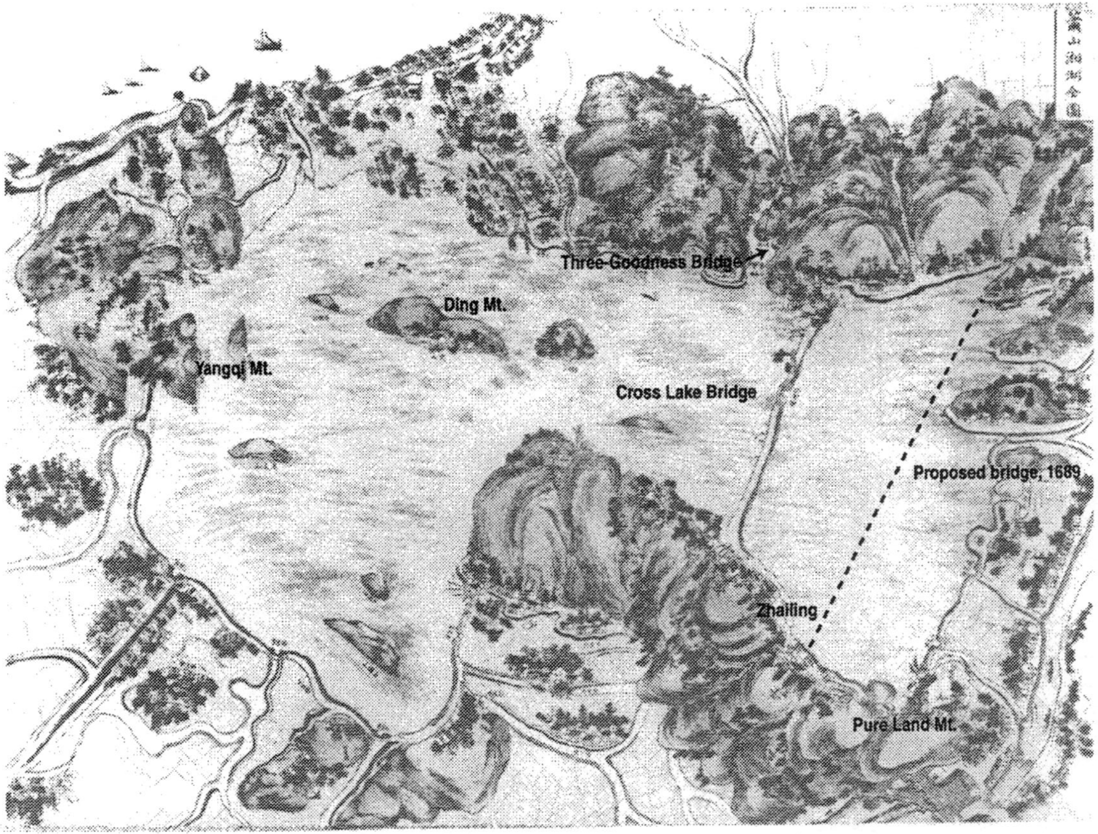

Map 8. Xiang Lake Sites, 1689

VIEW: *On Pure Land Mountain, 1689*

Near the county seat on the east side of the lake was Pure Land Mountain. At its base was Golden Spring Well, a source of water not even exhausted, it is said, at times of severe drought such as occurred in 1689. The spring was renowned for the purity and sweetness of its water, and multitudes of the drought-stricken clogged the roads and contended for access to the prized resource.[1] Like those who struggled for lake water, well users fought with one another to fulfill their own needs.

Located nearby was the temple that had given the mountain its name. Belonging to the popular Buddhist Pure Land sect, the temple was constructed early in the Song period, a century and a half before Xiang Lake became reality.[2] Destroyed in the early Ming period, the temple was rebuilt during the 1620s by a local notable, Cai Sanyue. Active in the reconstruction at the same time of the nearby Stone Pond Outlet sluicegate, Cai must have spent considerable funds in the six-columned temple; he also reportedly enlisted a monk to live there permanently as custodian. Associated with the temple was a pagoda where nightly lanterns were lit from dusk until dawn. Travelers on the Qiantang River and the sea reportedly could see the flickering lights on what became, in effect, a land-locked lighthouse. With the Xixing lighthouse, it became a guidepost for those navigating through the shifting shoals of the Qiantang and its estuary. It is perhaps not an exaggeration to suggest that, like the supernatural lights which appeared on the Qiantang's bank in the late

Yuan period to thwart a river crossing, the lanterns on the Pure Land Pagoda saved a number of lives.

Pure Land Mountain was near enough to the county seat to perform the same function as Chenxi Garden: an escape to nature from the responsibilities and humdrum of daily life. Wang Xianji, a jinshi degree holder of 1670, described the mountain and its temple as "a beautiful place" to which people walked from their houses in the cool of the evening.[3] The mountain, in fact, had its own garden, perhaps not very invitingly called Dripping Marsh Garden. If, like many Chinese, one sought a view from higher on the mountain, the pagoda or the Pavilion for Viewing the Lake offered good vantage points.[4] From there, Wang noted, one might watch the changing colors of the lake; the setting sun dyed the clouds and, through their reflections, the lake itself into continuously changing hues of orange, pink, rose, mauve, and blue. In such splendors of nature, one might indeed muse, as Wang did, on Buddhist spells at the approach of night.

Situated near the Pure Land Temple was the Shrine for the Virtuous and Kind, built in the middle of the fifteenth century to commemorate the creation of Xiang Lake by Yang Shi. By that time the earlier Shrine of the Four Senior Officials had fallen into ruins. At the new shrine, originally advocated by Wei Ji, spring and autumn sacrifices were instituted; in 1466 Prefect Peng, who had directed the Qiyan Mountain excavation, visited the shrine at its inauguration. His presence points to the significance of such shrines, which were common in most Chinese communities: the state supported their establishment to recognize local leaders who were models of public-spiritedness and self-sacrifice.[5] The transmogrification of exemplars of civic virtue into beings worthy of sacrificial rites underlines the effort of officials and local leaders to inculcate elite values to commoners and to bolster the importance of a particular exalted virtue with religious trappings. To a culture in which remembering the past was crucial for defining the present, such shrines provided illiterate and literate alike tangible structures of remembrance.

A generation after the construction of the shrine to remember Yang, Wei Ji himself was venerated there. Statues to represent Yang and Wei were erected in the main sacrificial hall, which was reached through entrance and ceremonial gates.[6] Close by was a pond, and south of this pond was a hall for sacrificing sheep and cattle. Ceremonies in the spring and autumn were designed to impress participants and onlookers with the ethical values of Yang and Wei—specifically, dedication to preserving the lake for the people of the nine townships. Such rituals reinvigorated the community with those values necessary to face new challenges to the lake. A clearing of ten mu (over an acre) to the west of the main hall was apparently used at times of the biannual rites.

An academy (*shuyuan*) located on the east side of the shrine complex was also built at the time of the shrine's construction. We are not told the use of the academy, though the establishment nearby of a kitchen and bathhouse suggests that it was more than simply a library; perhaps it was used as a neighborhood school for the young or as a place for scholars to meet for discussion.[7] Called the Southern Path Academy, it had two halls containing five rooms. Two tablets inside the academy were engraved with the names and titles of Yang Shi and his contemporary county military officer, Yu Dingfu, both of whom had reputations for the conscientious promotion of education.[8] At the shrine's spring and autumn rites, these tablets and the academy were incorporated into the ceremonies.

The Southern Path Academy predated the countrywide remarkable expansion in the number of academies in the sixteenth century. There is nothing to indicate that it became a distinguished scholarly center or that any noted scholar came to teach there. Some academies in the Ning-Shao Plain developed into highly significant centers of philosophical, historical, and literary discussion and transmission; some took on political roles in the closing decades of the Ming dynasty.[9] Yet the Southern Path Academy and the Shichuang Academy—the only other established on the shores of Xiang Lake—seem to have been in the backwater of such developments. The Shichuang Academy on Lian Mountain to the northwest of the lake was built in 1568 near the tomb of Huang Jiugao, noted West River dike preserver and author of the poem on the effects of Sun's Cross Lake Bridge for the future of Xiang Lake. While it is unsurprising, it is nevertheless noteworthy that the two academies in this water-dominated peninsula were linked to men whose career achievements were in water-control efforts.

In 1517 another change came to the Shrine for the Virtuous and Kind when the tablets for Censor He Shunbin and his filial son, He Jing, were placed on a stone table near the left column of the shrine. We do not know if some opposed the inclusion of the He father and son so soon after their deaths and with the waxing power of the Suns. That only tablets, not statues, were added may suggest some opposition. It is important to see that their inclusion came before Sun Xuegu's degree success and therefore probably incurred less opposition than if they had been added later. It is easy to understand the sentiment for including them—both had given their lives for the lake, Shunbin in a literal sense, Jing through his career. A poem by Lai Jizhi, a late Ming–early Qing dynasty scholar official, titled "He, the Filial Son" saw them as one spirit separated into two men whose names would remain fragrant with the incense of the respectful throughout history.[10] Lai's poem points to the literary transformation of ethical-political leaders to the religious realm, just as the shrine facilitated that change in a physical way.

With the strength of the Suns, it is not likely that much local emphasis was placed on ceremonies at the shrine. The Suns had precious little sentiment for venerating either the He or Wei family, who had opposed their ancestors; other local leaders showed little incentive to keep alive the values of the lake preservers or their shrine. Near the end of the Ming dynasty, the academy collapsed in ruins. The tablets of Yang and Yu were moved temporarily to the stone table that held the tablets of He Shunbin and He Jing. We do not know if the deteriorating shrine hall suffered physical damage during the violent change of dynasty in the 1640s, but sometime in the first reign period of the Qing dynasty (1644–1662), it fell to the ground. Someone rescued the two statues and the four tablets, and they were moved out to the side of the ceremonial gate.

In 1689, after thirty to forty years uncovered and at the mercy of nature, they were still there, probably hidden by plant growth and disintegrating along with the values of the men they represented. The abandonment and desolation at the area at the base of Pure Land Mountain was apparent. A look directly across the lake in the late summer of 1689 was enough to make the spirits of Yang, Wei, and the two Hes even more desolate than the site of their destroyed shrine: the Suns were building another cross-lake dike.

CHAPTER: *Image and Remembrance, 1644–1705*

Water, as part of nature, may take on many images, but these depend on water's particular state. If water rampages, as in the great floods from the Qiantang River in the summer of 1642,[1] man generally assigns it the image of destroyer. Put more generally, when nature through storm, flood, or other calamity erupts in violence, it destroys every image of itself except that of destroyer. In obvious contrast, the water in Xiang Lake and other bounded bodies of water can take on almost limitless images bound only by the imagination. Water, for example, is a reflector—of rising and setting sun, of moon, of clouds, all of these whole or fragmented into glistening bits of silver or shining gold flowers tossed in the waves.[2] Bao Bingde, writing in the early Qing period, could see "dark clouds reflecting like tears in the jade water,"[3] whereas Cai Zhongguang could see "clouds flying, reflected in the water like dispersed banners in disarray."[4] Imaging, like remembering, springs from the interplay of the imaginer, or rememberer, and the object.

Xiang Lake was not, of course, always seen as a reflector. People imposed other images depending on their relationship to the lake. Xiang Lake was a reservoir, storing water to pour forth in drought-destroying streams. Xiang Lake was a source of food, of fish and plants pulled from the depths. Xiang Lake was a transporter to gravesites, tree-shrouded pavilions, and crumbling shrines. Xiang Lake was a source of pleasure, intensifying the giddiness of a wine-induced reverie. If nature in its fury imposed the cataclysmic image, people imposed their own images on nature's placidity.

Nowhere is this clearer than in two poems that celebrate the lake harvest of the water-shield plant, the highly desired succulent plant for which the lake was renowned. The poets: Mao Qiling (1623–1716), chronicler of the lake and brilliant scholar, and Lai Jizhi, jinshi degree holder (1640) and official. Mao's images are traditional, straightforward, and culturally apposite; Lai's are surprisingly sensual, even sexual.

Mao:

> At dusk on a spring evening before Yewu Mountain, A-zi goes to gather the water-shield plants.
> The lake is filled with lotus root and leaves of the water chestnut.
> The small boat floats, heedless of its destination.
>
> A pole, ten feet long, pulls out the jade silk;
> The fragrance of the yielding water plants is like that of stirred-up flowers.
> Since people around the mountains are difficult to entertain,
> You must fill baskets to overflowing if you desire to send the plants to anyone.[5]

Highlighted here are the traditional picture of the lake's productivity, the quintessential time of day, if not season, for a lake excursion, and the fragrance and color of the plants, with the oft-used image of jade, as trite as the "kingfisher blue" attribute. The last two lines take us from the world of nature to that of society and the necessity of treating with care and propriety those to whom one is linked through social ties.

Lai:

> I want to gather lotus leaves; they can be made into clothes which even an inferior scissors can cut.
> I want to gather the flowers of rushes; they fly like snow into the horizon.
> I take a boat onto Xiang Lake; on all sides I am surrounded by mountains.
> In the lake are the precious jade grasses distinguished from all the other sweet-smelling plants,
> Planted amid the mists by the dragon in the depths of the water.
>
> It is like a beautiful virgin,
> Standing, not leaning, under the cold heavens,
> Looking at her reflection in the mirror of the lake.
> It is as though bright cold jade has become her body
> And her curling leaves do not yet disclose the luxuriance of her heart.

My emotions are overwhelmed thinking of her beauty.
I gather the plants in from here and there.
Under the jade heavens in a jade cauldron, I warm pieces of the plant in lake water.
Amid the mountains there is one who can prepare this dish;
But it is a delicacy which must not be readied in an ordinary way.[6]

After a rather uninspiring first stanza, Lai uses the senses of smell, sight, and taste to conjure a considerably different picture of gathering the water-shield. The male-female imagery of the second and third stanzas heightens the sensuality. Equally remarkable is the last stanza's thrust for personal satisfaction and pleasure, carrying none of the social import of Mao's piece. Here then are two views of the same action performed on the placid lake, but the images, indeed the interpretation, of those actions seem sharply dissimilar. The artist as interpreter imposes images on nature just as the rememberer imposes images on the world of men.

In 1645 and 1646, the lake and other local bodies of water assumed a sharply different image from the productive, life-bestowing reservoir with its treasures of plants and animals.

Mrs. Zhao, née Zhang, fearful of rape and death at the hands of the invading Manchus, drowned herself and her young son in Xiang Lake.

Mrs. Wang, née Dai, drowned herself in Xiang Lake.

Mrs. Lai, née Cheng, drowned herself in White Horse Lake.

Mrs. Lai, née He,
Mrs. Lai, née Huang,
Mrs. Lai, née Ren,
Mrs. Shen, née Lai,
in a suicide pact all threw themselves down a well west of Xiang Lake and drowned.[7]

Yang Shoucheng, his wife, née Tang, and their son threw themselves in Tiger Village Pond and drowned.[8]

Weng Sun and eighty-nine members of the Shen lineage drowned themselves in the Qiantang River.[9]

The lake, the river, the well became agents of death for the horror- and panic-stricken, to escape the fury of the Manchu army. The depths of the lake became the apartments of the dead, the corpses mingling with the tangle of water plants, then floating to the sunlit surface, where their stench, like nature's fury itself, destroyed every possible image of fragrance, leaving only the nauseating smell of death.

The mid-seventeenth-century crisis that led to such acts of desperation began with the collapse of the Ming dynasty amid large-scale rebellions led by ethnic Han Chinese. But the apparent victory of one rebel, Li Zicheng, in the spring of 1644 was quickly snatched away by defeat at the hands of Manchurian armies that entered China at exactly the right moment and seized power. The dilemma faced almost four centuries earlier when the Mongols drove the last Song emperor into the sea had reappeared: How should Chinese react to first the specter and then the reality of "barbarian" control? The choices of several men from the lake area—Lai Jizhi, Cai Zhongguang, and Mao Qiling—give further insight into the parameters and dynamics of China's culture and society.

The Lives of Lai, Cai and Mao before the Manchu Conquest

Lai Jizhi, of the noted lineage in the town of Long River west of Xiang Lake, was nicknamed Tang Lake, after the lake that his great-great-great-great-grandfather Lai Can had diked for irrigation in the mid-fifteenth century.[10] He received his jinshi degree in 1640, a year after gaining his juren degree. In the turmoil at the end of the Ming dynasty, Lai found himself in the capital of Anhui province on the staff of the imperial forces, which were assigned to crush the spreading rebellion of bandit-rebel Zhang Xianzhong. Lai was a quartermaster, and he quickly discovered that the morale of Ming forces was low. With impunity high-ranking officers frequently reduced their soldiers' rations in order to keep larger allotments for themselves. When Lai took over, he maintained order and raised morale by insisting that rations be distributed according to regulation.

In 1643 Lai's expertise at mediation was put to the test when one unit of imperial soldiers went on a rampage, plundering area inhabitants and causing many casualties. The root of the problem apparently lay in disagreements between two important officers. Effectively demonstrating his ability, Lai was able to smooth over the differences between them. In

1644, with first the Ming collapse before the onslaught of rebel Li Zicheng and then the Manchu invasion, Lai returned to Xiaoshan.

Cai Zhongguang, whose mother was from the distinguished Lai lineage, was a brilliant, obsessive scholar who pored over the classics night and day.[11] His biography notes that few could surpass him in intellectual ability. He attained the lowest-level civil service degree. Possessed of a retiring and rigid personality, Cai disliked change; he was often given to obstinate and eccentric behavior. He liked nothing better than to throw himself into the study of such classics as the *Book of Changes*, the *Analects*, the *Mencius*, and the *Spring and Autumn Annals*.

Mao Qiling was born in 1623, the youngest of four sons.[12] Both his grandfather and father had received honorary titles for their ability and virtue. A precocious child, Mao could read and understand subtle verbal distinctions by age four. He early on gained a reputation for his acerbic wit. His father reportedly was concerned that Mao's quick mind be channeled to produce proper behavior and not to become simply an instrument for virtuosity.[13] By age fourteen Mao had already passed the first level of the civil service examination system.

When the Ming dynasty collapsed in the spring of 1644, Mao, Cai, and two close friends from the county went to the mountains to the south, built a mud hut, and spent over a year in retreat, reading the books they had brought and engaging in intellectual discussions.[14] All in their late teens and early twenties, the men were so close that they were known at least in some county circles as the Four Friends.[15] Friendship, one of the five Confucian bonds, was highly significant in Chinese culture; in a highly hierarchical society it was the only essential social bond that was established between equals. Theoretically, at least, the barriers built by social differentiation in the hierarchical bonds could be avoided in the more relaxed relationship between friends. The bond took on added importance in a society that was based on arranged marriages and in which open, close male-female romantic relationships were generally not possible.

The Manchu Takeover, 1645–1646

For the people who lived around Xiang Lake, as for all those in Xiaoshan county, the years 1645 and 1646 brought the trials of living in a rapidly militarized area as an array of motley military leaders built defensive positions to defend against the Manchu threat. After their seizure of Beijing, it was a full year before Manchu forces reached Nanjing.[16] As they approached Hangzhou in early July 1645, three possibilities—collaboration, escape, or resistance—made up the range of reactions for those in Xiaoshan.

Many dismissed the collaborative route out of hand, fearful of what this barbarian rule might mean for their lives and property or out of loyalty (obligation) to the Ming regime under which China had lived for almost three centuries. In the tense situation, those men who continued to work as county officials for the new Manchu regime were called traitors; several were seized and slain summarily in July 1645.[17] Escape seemed an inviting route to many. Whatever their motives, those who chose suicide, it might be said, tritely, selected the ultimate escape. For such elites as the Four Friends, escape might be to a bucolic mountain retreat where hermitlike they might remain until the turmoil passed. For most, escape meant traumatic flight, leaving home and most possessions to venture to an area less likely to experience the ravages of war. Becoming a refugee, however, was not easy. It cast the already psychologically and socially uprooted into dangerous and unpredictable situations. Away from home, how would one live? Away from home, how would one speak to villagers whose dialect (so different it might as well be a foreign tongue) might differ from village to village? Away from home in a culture in which outsiders were considered as threats or as nonpeople to be ignored, how would one cope? Only staring into the face of the specter of rape, pillage, and murder could the choice of flight become feasible. Many did not, perhaps could not, flee and hoped against hope that they could avoid the battle or escape the marauding bands of soldiers who had scant regard for the lives or property of these people who spoke a different language and cultivated different customs. Many stayed to farm, fish, dig mud, carry on trade, and live as though the world around them were peaceful, ignoring the turmoil if it did not affect their own personal interests.

In the summer of 1645, resistance increasingly seemed a promising option. When Nanjing fell, Zhejiang men who were loyal to the Ming began to press Zhu Yihai, a prince of the Ming royal household who had taken residence south of Ningbo on the Zhejiang coast to become "administrator of the realm" and serve as rallying point for the resistance.[18] When he agreed on August 19, he made Shaoxing his headquarters. For the first time since the twelfth and thirteenth centuries, when Hangzhou had served as imperial capital, Xiaoshan found itself close to an imperial court, even if one of a pretender. That proximity only increased Xiaoshan's danger, especially as the Manchu threat massed at Hangzhou across the river: Xiaoshan lay between the Manchus and their target at Shaoxing.

Even before the temporary court at Shaoxing was established, many in Xiaoshan had made their decision. Brothers Xu Fangsheng and Xu Fanglie and their fellow students had met in the symbolically important Confucian temple in Xiaoshan to swear tearfully their loyalty to the Ming; Xu Fanglie reported that 113 men promised not to serve the new Qing dynasty.[19] The Xu brothers were friends of Mao Qiling and Cai Zhongguang.[20] As individuals decided to resist, five major area military leaders raised five to six hundred troops each; the wealthy were to provision these troops through contributions. Xu Fanglie actively contributed, as did men like Yu Deyuan, a rich merchant of Wen Family Dike on the Qiantang River. Enlisted soldiers were garrisoned in key defense spots: Xixing, the most crucial, at the mouth of the Qiantang across from Hangzhou; Tantou on the Qiantang, west of Upper Xiang Lake; Fish Lake and Yiqiao on the Puyang River south of Xiang Lake; Long River, the home of the Lai lineage at the foot of Crown Mountain; and Qitiaosha, along the Qiantang south of Xixing.

In addition to the garrisons, local notables enlisted their own "righteous soldiers" to be stationed at key points. Lai Jizhi, returned from his military provisioning post in Anhui, chose to aid the resistance by raising troops from the White Horse Lake region with a kinsman. The Lais bolstered the garrison at Qitiaosha. Two leaders of the Zhu lineage from east of Xiang Lake led their own "righteous soldiers" to hold the strategically important Turtle's Son Gate on Xiaoshan's Hangzhou Bay coastline.[21]

The initially promising resistance had been bolstered in early August by an episode that echoed the past. The head of defense efforts in Xiaoshan, former prefect Yu Ying, became aware that more than a hundred Qing boats and many hastily constructed rafts were being readied for invasion. On the night of August 5, he dispatched a suicide squad to destroy the rafts and sink the boats. That same night a strong wind and great tide sprang up, dashing apart the boats and rafts. It seemed to many that the spirit of Li Di had once again saved the county.[22] Even though the defensive strategy seemed to work throughout the fall of 1645—climaxing in the pushing of all Manchu forces west of the Qiantang River on December 1—there were signs that the situation was deteriorating.

The "administrator of the realm" seemed to be a poor judge of loyal subordinates. After the establishment of the court at Shaoxing in late September, he called two notorious Ming officials, Fang Guo'an and Ma Shiying, to direct the military defense; both were known for their self-aggrandizing opportunism. Although Ma never reached the area, Fang put his nephews, of questionable military ability, into strategically crucial leadership positions. By late October, military commanders could agree neither on overall strategy nor on methods of coordination.[23] If the leaders paid any attention to omens, what happened at the October 16 sacrifices to the Qiantang River (which had "saved" them in August) should have been seen as fateful. The sedan chair of the general performing the ritual was dropped accidentally; items to be used in the ceremony and some of the general's ceremonial clothing were damaged or ruined.[24]

Most important, however, was the growing disaffection of the populace. The military had become omnipresent. The people were continually expected to provide more soldiers and send more provisions. Assaulted by the seemingly inevitable military plundering—some of the worst of which occurred around Xiang Lake and Long River near Fang Guo'an's garrison—the populace saw the confiscation of the wealth first of those who had thrown in their lot with the Manchus, then of the very wealthy Ming supporters, and finally of the commoners themselves. Military demands were insatiable; officials sent to local posts were reported to be venal and crafty.[25] Finally, the military commanders fell to fighting among themselves for provisions. Fang Guo'an seized for his own use funds that had been intended for all the military commanders; morale deteriorated as disharmony mounted.[26] In late December 1645 or early January 1646, an altar was built on Crown Mountain, base of the Lai lineage, to do obeisance to Fang Guo'an as the supreme general (*dajiang*).[27]

Sometime in late 1645 or early 1646, Mao Qiling, eschewing escape, left the retreat he had shared with Cai and their friends, opting instead to resist. We are not told how his friends reacted to his decision; we do know that he and Cai remained close friends.[28] Whether Mao's decision reveals that his father's instructions to Mao's early teacher about the importance of Mao's understanding moral values had borne fruit or whether this seemed a good opportunity for Mao to further his career we are not told. His uncle Mao Baoding was active in the resistance; perhaps through his advice, Mao Qiling was recommended to the "administrator of the realm" as a man of genius who could help in the better administration of the military effort.[29] Mao first demurred at the offer of a post, but he then agreed and was assigned to Fang Guo'an's army. Immediately repelled by Fang's policies and attitudes, Mao refused to keep his views hidden. When he was invited to welcome the arrival of the infamous Ma Shiying to Xixing, he refused, telling his uncle within earshot of Fang, "Fang and Ma are traitors. The prince is now trying to raise a righteous banner. How is that possible working together with two thieves?" Mao then told his uncle he wanted out of his post. Fang was furious with Mao, and Mao resigned shortly thereafter. Morally incensed at the nature of the Ming loyalists, Mao again embarked on an escape route, this one more permanent: he retired to a mountain monastery where he donned the robes and accepted the tonsure of a Buddhist monk.[30]

In the summer of 1646 the Manchus massed many troops on the Hangzhou side of the Qiantang, having eliminated resistance in Anhui and southwestern Zhejiang. At the crucial moment in early July, Fang Guo'an, the supreme general, abandoned his garrison to flee with the "administrator of the realm" to Taizhou on the east coast. The Manchus easily crossed the river due to an extraordinary drought that had left the Grand Canal completely dry and allowed them to wade across the Qiantang.[31] Then the people of Xiang Lake, objects of military extortion for a year, felt the brutality of Manchu military might. Women drowned themselves to escape rape and murder. Lai Jizhi's brother-in-law, Yang Yunmen, chose death at his own hands rather than suffer torture by a northern barbarian.[32] Yu Deyuan, provisioner of the aborted resistance effort, stabbed himself to death.[33] The brothers Xu and Cai Zhongguang retreated from the world as hermits; Cai even refused for a time to talk to anyone.[34] With the thousands of young who died in the resistance, the suicides of the able, and the political eremitism of the once-committed and brilliant, the generation that came of age in the 1630s and 1640s was truly "lost."

Even those who withdrew from the mainstream of life could not undo through their actions the traumatic victory of the Manchus and the fact of Manchu control. The first decades after the conquest were necessarily a period for assimilating what had happened: the horrifying vengeful atrocities; the deaths of friends and family members; the immense task of reconstruction; the imposition of the demeaning symbol of subservience to the outsiders—the queue; and the early, tentative attempts of the conquerors to reach out to the conquered.[35] Just as people had reacted differently to the conquest, so they responded to its aftermath in varied ways.

Mao Qiling remained at the monastery about five years, after which time he forsook his Buddhist life and returned to the life of a scholar with the lower-level degree he had attained in 1637.[36] Of Lai Jizhi's activity we are told little immediately after his leading "righteous soldiers" during the resistance. We do know that the youngest of his four sons, Yanwen, was born probably in the 1650s, that, therefore, at least his home life had probably maintained or reacquired a sense of normality.[37] Cai Zhongguang continued his eremitic retreat until 1681, first concentrating on his study of Ming dynasty calamities and portents (perhaps commenting on the omen of the oranges a century earlier) and later focusing on the ancient Xia and Zhou dynasties. In his spatial hermitage he retreated farther and farther into the past.

The Xiang Lake Poetry of Lai Jizhi: A World Reconciled

The Xiang Lake poetry of Lai and Cai—of Mao we will speak at length later—provides excellent pictures of the lake and its people; even more, it offers insights into sharply different reactions of these men to life after the conquest, suggesting the range of responses among intellectuals in the region. Lai clearly reconciled himself to life under the Manchus, and, in spite of repeated references to his worries and anxieties, his poetry has a predominant mood of contentment and acceptance of the world as it had come to be. He composed a cycle of four poems on his visits to Buddhist temples on Yangqi Mountain on the lake's southwestern shore.[38] Even though the poetry makes clear Lai's appreciation of the beauties of nature, it is rather his ties to Buddhist elements, not the beneficence of nature, that bring him solace. Nature and religion are the prime elements in his poem "Passing by Yangqi Temple."[39]

> I have come again to Yangqi Temple:
> Everyday cares, excessive desires turn completely to ashes.
> This stream will never lure a fisherman;
> The clouds on the mountains will open only for a pure heart.
> In the middle of the day, the monk entered his hut to worship.
> In the middle of the night, the wind produced ten thousand claps of thunder.
> I read yesterday's poem today:
> My self-knowledge swallows up my self-doubt.

Floating from afar, a skiff has arrived.
Carnal thoughts are half-scattered; principled thoughts grow.
On the rocks are shadows of the swaying pine trees.
The red stairs before the split crabapple tree [lead to] the old Buddha.
Whenever the lantern's lit, I look for a living sentence:
The former dynasty as an old matter worshiped (Yang) Qi Wang.
I would like to follow the fisherman as a vagabond
And see for ten thousand li the vastness of nature.

A key sentence in this poem—"The former dynasty as an old matter worshipped (Yang) Qi Wang"—suggests Lai's acceptance of the new. Yang Qi Wang was a Song dynasty figure who was buried, it was traditionally thought, on the mountain. Mao Qiling had argued persuasively that this was not the case, and Lai seems to be willing to impute this practice of worship to the old matters of a former dynasty.[40] Furthermore, the quenching of carnal desires and burdensome anxiety through turning to Buddhism leads Lai to an acceptance not only of himself ("my self-knowledge swallows up my self-doubt") but also of nature.

Carrying further the themes of acceptance of the present changes of the world and the positive relationship between man and nature is Lai's poem "Passing by Yangqi Mountain's Chongfu Temple." Chongfu Temple was dedicated to Yang Qi Wang (here called Duke Wang).[41]

Around the many villages rises the cooking smoke.
When I arrived at the mountain, I cast off my many worries.
The water [Xiang Lake] from the mountain cascades, stretching far, is good for the spring tea;
The bamboo shoots, sprouting forth, provide fresh vegetables for the evening meal.
Duke Wang from former times has a vacant grave.
Because of the beauty of the scenery, birds and fish come, disturbing my contemplation.
No one speaks of goings-on in previous years on the river;
And in mountain valleys, there are not yet any cultivated fields.

Why is Jiangnan completely committed to mourning?
In the mountains, there is still a place for meditation.
The tranquil bridge crosses the water where all kinds of fish live together.
The bamboo plaiting protects people's flowers, but flowers plant themselves.
Beside the pagoda, there are the ruins of an old inscription with names of guests from of old.
The Buddhist lantern has already come to light the evening.

> The evening clouds generally come before the gate as a lock,
> Impartially permitting neither the people at leisure nor those who are busy to open it.

The last lines of the first stanza and the first line of the second point to the aftereffects of the conquest in Jiangnan (the area south of the Yangzi River): the devastation of the farm land and silence about the former defense against the Manchus. Yet Lai's view is unremittingly optimistic and conciliatory: the productivity of the natural world (the water, tea, and bamboo shoots), the solace of nature (a place for meditation dispelling his worries), and nature's universality (the fish and flower images) all point to the uselessness of mourning about the conquest. Nature, which provides the cloud-lock at the gate, is impartial, allowing neither "people at leisure nor those who are busy" to enter.

In three stanzas he wrote on Xiang Lake, Lai points to the lake itself as a retreat from the world.[42] He notes the busy pace of life in the area, contrasting it with the solitude and beauty of the lake.

> On the road to Xiling people are like ants;
> On the Raksha river [Qiantang] bank, the waves seem like mountains.
> But there is only a suspended fishnet in a corner of the lake.
> The world of nature is always peaceful.

The final stanza juxtaposes the world of nature with such historical sites around the lake as the King of Yue Garrison. This contrast does not bring to Lai's mind, as it does to many Xiang Lake poets, the insignificance of humanity and human works before the awesome continuity of nature. Instead Lai describes the historical sites surrounded by the productive luxuriance of nature—trees with blossoms like the phoenix, bees making honey, the bamboo, and plentiful fish.

> The mountains all around are like a curtain without seams
> At the foot of which is the jade emptiness.

Lai's vision of life includes the interconnectedness of all things ("a curtain without seams") and the synergism of man and nature. His poetry reveals a man reconciled to life after the fall of the Ming, seeing a world (nature and humanity) existing in harmony. There is no melancholy here, and there is not a single mention of wine, for most Chinese poets an agent of escape. Not once does Lai image nature in other than its placid state.

Lai did not spend his life after the conquest in peaceful retreat. Sometimes in the late seventeenth century, with his son Yanwen and his kinsman Ersheng, he became active in local fiscal reform to relieve the undue burden of taxation on poor households.[43] In the sixteenth century the Ming state, under a fiscal policy known as the Single-Whip system, had attempted to merge various taxes and convert the payment of in-kind land taxes and labor service taxes (called the head tax) to cash. This commutation, however, was carried out on the local level somewhat arbitrarily and left a system of inequality in which the head tax fell on the poorer households regressively. The Lais became active in efforts to abolish the excessive head tax and to have taxation computed only according to the land tax.[44] Jizhi attempted, in the language of his world view, to reconcile the tax system with the economic realities of life around the lake in order to soothe the lives of the poor.

The Xiang Lake Poetry of Cai Zhongguang

In contrast to Lai's acceptance of the conquest, Cai Zhongguang's attitude about what had happened at midcentury was bitter. His life as recluse stemmed not from misanthropy but rather from the specific horrors visited on the area by first the defenders of the Ming and then the Manchus. In a sardonic attack on "the imposing, warlike general" Fang Guo'an, Cai uses symbols of power—warships, banners hung on willow trees at Fang's garrison west of Xiang Lake, even Fang's hirsute arms (like a mane)—to suggest Fang's catastrophic effect on the area.[45] Interested mainly in his own aggrandizement ("booty" in Cai's poem), Fang had fled his post at a crucial time in the Manchu drive, and Cai aptly notes that this bête noire of Mao Qiling brought only turmoil and impoverishment to the area east of the Qiantang. Part of Cai's antagonism to Fang undoubtedly reflected the views of his close friend Mao, but there seems a consensus that Fang sought primarily private gains at a time of great public crisis.

In almost every way Cai's poetry reflects a different view of the world from Lai's.[46] Although nature can be peaceful and may help empty one's heart of anxieties, nature itself often brings "melancholy" and "excessive sorrow."[47] Amid a retreat into peaceful nature, he wonders why his heart was still filled with disquietude.[48] Furthermore, nature is not always placid: the "west wind blows wildly"; "wild clouds" scud across the sky; the wind destroys the flowers as their petals are blown into the mud, grass, and water; an earthquake shakes the King of Yue Garrison; the lake and sea have a bleak, chilly appearance; and the Qiantang has whirling waters.[49] The strong sense of inner agitation seems to be connected to the

Manchu conquest. In a poem on spending the night at a peasant's home on a Xiang Lake island, he talks of the simple life of the peasant in the context of what has occurred.[50]

> Formerly this land was hidden next to the mountains.
> (When it was filled with weapons, my heart sneered.)
> A sail and mast when crossing the lake are suspended over the water, casting small reflections.
> Recently it was taken possession of by saddle and the Manchu whistle.
> Today the grass along the road is withered.
> There is the soughing of the cool wind in the trees under the moon.
> Beside the embankment the chickens and pigs are fed after the field work is done.
> After the oxen are fed inside the bamboo door by the maples,
> Then is the time [to relax], buy wine, wear beautiful clothes,
> To become a gentleman—and have a song full of tears.
>
> The host in this world of nature is the old man of the fields.
> The guest [Cai] is seated. Who plays the five-stringed lute?
> Many things happen to people that they cannot talk about;
> In the past the host and I shared the damming up of our feelings;
> However we'll take the hoe to plow up the barren southern mu.
> In the middle of the night, unable to sleep, I dance with swords;
> Before my eyes the flowers are blown against the earth;
> In the silence wild ducks call,
> Before the lamp light, war carriages turned, hastening the attack;
> As they crossed the river with the clouds against the sand, the river washed over their metal armor.
> I heave a great sigh over that time, now become peaceful,
> Desiring to share this view with my host.
> The passage of time does not affect people living in nature.
> Night after night the moon over the mountain village is bright.

In another poem on the lake Cai notes that wars and revolution bring deserted villages, and his inner turmoil, in spite of what might be called a veneer of Daoism, belies a sensitive intellectual alienated by the events of his time. In five of his thirteen extant poems on the lake, separation is a key theme—separation from friends, nature, and the past world; it is often symbolized by the broad expanse of the Qiantang, which separates Xiaoshan from land to the west.[51] It is perhaps not surprising that most of Cai's poems make reference to wine and getting drunk; though many of his poems are written for friends or are accounts of outings with friends, his imbibing is not convivial but escapist.

Whereas Lai linked nature in a positive way to the passage of time and history, Cai saw them as not akin at all. In the above poem people in nature are not influenced by time. In a poem about an outing on the lake with friends, he makes clear his view that the world of nature (*fengjing*) is separated from the "orderly arranging of years" by man and that, although he is in the world of lake reflections, mountains, and flying egrets and orioles, he still felt hatred for the King of Yue—the historical symbol of power.[52]

Cai's outlook is not completely bitter; he is aware of nature's beneficence and of others around him, both friends and peasants, noting them far more frequently than Lai, whose only poetic characters are lakeside Buddhist monks. In an unusual short poem, Cai points to the area's natural abundance for its denizen commoners and once again contrasts their contentment in the rustic life with his own disquiet and longing for escape.[53] He recounts an episode that occurred in the mountains around Xiang Lake: a deer, drinking at a stream, was caught and killed. The people feasted joyously on venison and wine.

> From where did the food come to stock the kitchen?
> It was caught and dragged by the feet,
> Then cooked, a delicious flavor.
> My wandering thoughts are like chasing the deer, running wherever it will.
> Watery mists cover the sun far away;
> Clouds touch the blue mountains.
> To let go, drinking with friends amid mountain and lake
> Relaxes my anxious thoughts; drunk, I write this poem.

Both Lai and Cai use the geography and setting of Xiang Lake to reveal the state of their minds following the conquest. Their images differ sharply, varying with their views of nature and history. In 1681 the Manchu emperor issued a special imperial mandate to recluse scholars to teach; thus, over three decades after the conquest, Cai and fellow recluse Xu Fangsheng emerged to expand on their studies of ancient history. Perhaps time had brought Cai some sense of reconciliation.

Mao Qiling, Chronicler of Xiang Lake

The imager refracts the image of an objective reality through the mind, which has its own interests and predilections, to create a new image: Cai's and Lai's pictures of the world of Xiang Lake and Mao's and Lai's images of the process of gathering water-shield plants are cases in point. History can be described as imaged rememberings from the past. Historical information or data (the rememberings) are actively shaped and interpreted (imaged) through the mind of the historian. We must understand that no text speaks without an imager and that knowing the imager-historian's own predilections provides an essential perspective on his work. Mao Qiling's account of Xiang Lake until about 1700 provides us with most of what we know about the struggle over the lake to that point. An account based on documentary sources, it has, like any history, been refracted through the mind of the author. A look at Mao's life and at evidence of his personality and views, therefore, provides insights into the nature of his history, a perspective important because Mao's images help shape our images of Xiang Lake's past.

In the 1650s Mao had a number of experiences with other figures in the county that probably contributed to his view of the lake area and its leaders.[54] As part of the process of joining a literary society, Mao compiled an anthology of area literati poetry. He selected four poems from a late Guiji county jinshi, Wang Zizhao, who had served in north China during disturbances at the end of the Ming. Wang at one point had to hide from bandits, and he did not return until his safety was insured.[55] Mao named that section of the anthology "Returning from Amid the Bandits" (*cong zei zhonggui*). Wang's father was enraged, feeling that Mao had purposely insinuated that his son had been involved with bandits. Mao cannot have been unaware of how this title would be interpreted; given his personality, one cannot help but suspect intentional sarcasm.

Mao seemed almost determined to offend local literati with his blatant opposition to the new dynasty (refusing to take further examinations in a Manchu-sponsored civil service system) and his literary activities. He selected, for example, as the subject matter of a drama an undistinguished Yuan dynasty figure who sold his daughter and gave himself over to a life of theft. Local literary figures pounced on Mao, accusing him of undermining the morals of the community and of throwing the role of the scholar into disrepute. Because of the ethical-political role of the educated elite in Chinese culture, the writings of the elite, whether openly political or not, frequently had political import. Mao was accused of encouraging immorality by his focus on thievery and of furtively charging those who had submitted to the Qing dynasty of "selling their daughters." He was arrested but was released by the prefect, who argued that there was no case against him.

Mao's arrogance and his disdain for his opponents only brought him more trouble. In the late 1650s he was framed in a murder case by people bent on destroying him. Accused of killing his wife's relative, with whom he had had an altercation on the road from Xiaoshan to Xixing, he had to flee the county and change his name. He was warned by Cai Zhongguang, "The anger is deep. Don't go out as you can't avoid being pointed out."[56] His name was finally cleared in the late 1660s. From his perspective, local bullies, engorged with their own sense of power and self-importance, had tried to ruin him to the point of framing him with a murder they had commissioned. The general outline of these events have intimations of the treatment of Censor He by local leaders almost two centuries earlier. There is little question that Mao too saw the similarities in the early 1680s when, finally acquiescing to Manchu rule and as a member of the board compiling the official Ming dynasty history, he was assigned to collect and edit biographies of key figures living in the period 1488 to 1522.[57] Here he undoubtedly first came upon the story of the Hes and their treatment by the Suns, the Wus, their hangers-on, and local officials. But it was probably Mao's own involvement with the Suns at Xiang Lake in 1689–1690 that led him finally to write the history and that provided the ultimate point of view of Mao's account of the lake.

Mao Qiling and Xiang Lake: The Crisis of 1689–1690

The climactic attempt of the Sun lineage leaders to turn a public reservoir into a private preserve came in the early fall of 1689. By the end of September, a summer drought had become so severe that the bottom of Lower Xiang Lake was completely dry and cracked. Like his ancestors, Sun Kaichen was ready and able to take advantage of opportunities for his own benefit.[58] Still exhibiting the power that certain leaders of the lineage had held for almost three centuries, Sun brought together several thousand people and constructed an east-west dike across Lower Xiang Lake from Zhihu Peak in the west to Zhailing, the traditional home of the Wu lineage on the east. The dike cut Lower Xiang Lake in two. Though the Suns could easily reach the county seat via the Cross Lake Bridge to the south or via the north side of the lake, the new dike, running parallel to Cross Lake Bridge, was simply a private lineage road for greater convenience to the county seat.

This drastic restructuring of the lake was reported immediately to the county yamen by a local *baojia* head. From the beginning both local leaders and commoners defended Sun, arguing that a Buddhist monk had prompted the construction on the pretext of convenience of travel to a temple. Although the magistrate responded to the baojia head's report with a call for public discussion between officials and the "people" (*min*), no one came to discuss at the announced time. The magistrate did transmit the report to the new prefect, Li Duo, who had received the governor's specific instructions to rectify local water conservancy problems in Shaoxing prefecture. On October 19 Li began an investigation of the complaint but immediately met the silence of the "wealthy families and powerful lineages" who refused to implicate the Suns. The power of the Suns and their allies was apparently so strong that the townships' elders (*lilao*) would not even discuss the issues. When Jiang Bangrui and Chen Dakui, two men from Lao Lake Village to the northeast of the county seat, finally came forward to support the original complaint, they and the investigating officials were met with petitions trumped up by the Suns and their allies accusing their accusers of being the actual violators of the lake. The cacophony of mutual accusations of the wealthy local leaders began to reach new heights of rancor, and the investigating official denounced the querulousness of all the elites (*tuhao*).

Mao Qiling, suffering from rheumatism, had retired to Hangzhou in 1686. By then a distinguished scholar-official, he was frequently in the attendance of the Zhejiang governor, who, after the news of the turmoil over the lake, consulted Mao the native son. The governor assigned Mao to investigate the situation. News of his assignment apparently spread quickly; when Mao crossed the river on November 18, he was met by a great crowd of local leaders clamoring about his sedan chair with legal complaint upon legal complaint—all labeled by Mao as "lies." The strategy of the Suns and their allies seemed to be to confuse the issue with an endless stream of accusations. Mao dutifully took down the names of the plaintiffs and proceeded to Shaoxing, delivering the list to the prefect. He inspected Xiang Lake, reporting that what he found and what the Suns and their allies had said did not match. Disgusted with the deceit of the Sun clique, he called for an immediate meeting of officials and people to discuss the problem. But even now, with this special emissary from the governor, there was only silence: no one would discuss the case. Mao returned briefly to Hangzhou for medical treatment but tried a few days later to sponsor a meeting; again no one came. A second attempt brought people together, but they simply sat and stared: those who knew about lake matters dared not set forth formal legal complaints. Even though the Sun's actions were endangering the livelihood of those irrigating from

Lower Xiang Lake and ultimately would affect all irrigators, area people feared immediate retaliation by the overreaching Sun clique, whose arrogance before the community and officialdom apparently knew no limits.

Frustrated and disgusted, Mao crossed the Qiantang to stay on the outskirts of Hangzhou away from the poisoned air of Xiaoshan; such a retreat would ostensibly allow him to clarify the problem in his mind. From an anecdote he relates, it seems more likely that, like many times in his career when he lacked the determination to continue a line of action, he was ready to give up and allow the Suns a victory. He must have been reminded of three decades earlier, when he was hounded from the county on the fabricated murder charge. Like He two centuries earlier, he seemed no match for the local "lords" and their supporters advancing their private interests. That evening he was approached by one Zhou Xuan, apparently the son of a friend. Zhou prodded him: "Don't you know about matters concerning Xiang Lake? The people of the county are looking to you yet you slight them as if they were a wife." The remark struck home, and Mao decided to recross the river to search out the truth and deal with the problems.

After his investigation, even though none of the local leaders would yet cooperate, Mao made known the names of the guilty, and he drew up what became known as the "four damages" and the "five impermissibles." Forwarded by the county government to Shaoxing, Mao's recommendations were ratified by the prefect. Mao's report specified the four damages brought by the Suns through the Cross Lake Bridge and the new dike: they blocked the flow of water; they contributed to the ecological disequilibrium of the lake; they cut water off from areas that needed it or were entitled to it according to the old regulations; and they destroyed Lower Xiang Lake. The last point was the key, for it was apparent that the Suns were making the lake their own. Because of inadequate drainage, Lower Xiang Lake had become clogged with waterweeds. The Suns sought more land there for kilns; they dug deep holes for tile-making clay; they fished in the deep holes; they gathered decomposed grass for their own fields' fertilizer; and they planted bamboos as boundaries for their own benefit, further slowing the waterflow. Mao made it clear that two lineages' private goals (*siyi*) could not provide the standard for the nine townships.

Thus he issued the "five impermissibles": rich lineages must not usurp lake land for their own uses; the color boundary of Guo Yuanming must not be broached (as it was, the Sun lineage lived mostly in original lake land); the disregard of lake regulations since the mid-sixteenth century must stop (Mao pointed to people near Ding Mountain in Upper Xiang Lake who had begun to carry in soil to make their own dike in the lake and to the Three Goodness Bridge, constructed by Lai Sixing across a corner of the lake in 1628);[59] appeals to the fengshui of graves located in the lake where they should not be must not be permitted; and the rules of honorable men of the past must not be destroyed.

The Suns and their clique resisted. From late 1689 to the spring of 1690, rain broke the drought, and the dike-builders claimed that the new dike made little difference. When the prefect, under Mao's report, gave orders to destroy the dike and punish the guilty with an immediate thirty bamboo floggings and time in cangue and fetters, the Suns bribed the yamen runners not to carry out the orders. The detailed magistrate's report, which restated Mao's, was sent to the prefect in October 1690. Not until the provincial judge and the treasurer backed the prefect in his orders, however, were the dike destroyed and Sun Kaichen and a Buddhist monk named Cui Hong punished.

The sources are silent on any violence that may have erupted in curbing those who for so long had done as they pleased; it is hard to imagine that no hostility ensued and that the Sun clique acceded willingly. We know that the Suns never again threatened the lake, though we do not know how their power was broken. *Official* persistence and determination to deal decisively with the problem for the first time since Wei Ji's attempt was certainly the key. It was also undoubtedly important that those opposed to the Sun power knew that the Suns would now be unable to retaliate and intimidate. Mao reported that the elders (*fulao*) of the area gnashed their teeth in hatred of the Cross Lake Bridge yet were afraid to incur the wrath of powerholders. The passivity of the Chinese around Xiang Lake may be charged to intimidation and fear, but it points even more to the immobility that stemmed from submission to authority, the trait inculcated in family training and through the various government ethical-political strategies that promoted officially prescribed virtue.

A word must be given to Mao the historian, the imagemaker. His frustrating involvement with the strongmen of the area probably gave rise to his arrangement of the history of the lake as a series of struggles. His images have become our images. Yet other sources—Zhang Mou's and Wei Ji's, for example—and the subsequent history of the lake corroborate Mao's interpretation that struggle was a major theme of Xiang Lake's history. A contemporary of Mao, Ren Chendan, wrote an account of the 1689 crisis that confirmed Mao's account. It seems likely, though, that Mao's vision of Xiang Lake's history is refracted through his mind to give inordinate emphasis to the tragedy of Censor He and his son. Both He and Mao had fought the Sun lineage, and for Mao, He was surely a tragic hero. Not only did Mao compile He's biography for the Ming history but the He tragedy became central to his account of Xiang Lake. He also wrote a lengthy account about the Hes' inclusion in a shrine. Mao must have felt a kinship to He: they had similar personalities—forceful, obstinate, certain of their rightness (though Mao may have been more easily discouraged). Their treatment by area elites was also similar perhaps in part because their arrogant approach gave rise to common reactions among the people.

Were the Suns overemphasized villains? Since the Wu lineage also shared every major violation of the lake, why is their role downplayed? The construction of the Three Goodness Bridge by an illustrious member of the Lai lineage—which added to the problems of Xiang Lake—is mentioned as if in passing. Though Mao's account is generally accurate, the images we have bear the nuances and images of the lake's history that Mao formed through his own personality, sense of identity, and biases. While Lai Jizhi saw gathering the water-shield plant in sexual terms, Mao viewed it in terms of social results. A lake history written by Sun Xuesi, Zou Lu, or Sun Kaichen might provide similar clashing images; certainly their reflections would heighten our awareness of the complexities and subtleties of the struggle over Xiang Lake.

Mao Qiling and the Shrine for the Virtuous and Kind

Mao's empathy with He Shunbin became even clearer in an episode of 1704–1705 regarding the Shrine for the Virtuous and Kind. The shrine lay in ruins when in 1703 the provincial governor presented the county magistrate over two hundred taels to be used as county leaders saw fit. They decided to use part of the money to rehabilitate the shrine.[60] At a meeting on June 13, 1704, 257 county leaders discussed not only the repair procedure but also a new issue: Should the tablets of the Hes be included in the shrine?

At first one might suppose that the Suns, as with the writing of the county's history, were trying to alter, for those yet unborn, the rememberings of the lineage record at the lake. Though they may have played a role behind the scenes, there is no intimation that they were involved. The record of the affair is Mao's, and it is hard to imagine his not including them (given his dislike of the lineage), even if he merely suspected their involvement. The primary opponents were the descendants of Wei Ji, keepers of the shrine. They opposed the incorporation of the tablets of He Shunbin and He Jing in the shrine because, they argued, the Hes' contributions to the lake did not match those of Yang, who had created it, and of Wei, who had restored it. The incorporation of the tablets was an affront to the memories of Yang and Wei. The Wei family brought forth formal technical objections and raised the issue of whether sufficient money existed from the land whose proceeds funded the shrine and its sacrifices.[61] The money issue may have been significant; critics of the Weis later charged that the lineage was establishing its own temple (*miao*) and was trying to use funds from the shrine lands.

The supporters of incorporating the tablets, chief among them Mao, overcame the arguments of the Wei spokesmen. They pointed out that the Hes had given their lives to restore the lake and that the tablets had rested at the shrine for almost two centuries. When the shrine was repaired, Mao Qiling, now in his eighties, helped erect the tablets of Censor He and his son at the side of the main shrine hall and participated in the fall sacrifices. As a result, in early 1705, Wei Qixian, the shrine's keeper, accused Mao and his supporters of usurping the sacrifices, shrine, and shrine property and called for another meeting of the area's elite. An account of the episode by the magistrate, dated April 24, 1705, however, indicates that the matter was closed.

The episode reveals Mao's dedication to the memory of He Shunbin; beyond that, it points to significant elements of Chinese culture. The importance of remembering the past infused the rebuilding of the shrine and the debate over the tablets. For the Chinese, the written record of the past—official history or county history—is crucial: in his long list of errors of the Wei lineage, Mao made frequent reference to what had been recorded in county histories since the early sixteenth century (including an account of the Suns' exclusion of the Hes' role from the late Ming history). Mao's detailed recording of the 1704–1705 episode indicates his sense of importance of a written account. But remembering goes beyond the written record to the shrine itself, a physical reminder of the ethical-political values to be cherished. Whether the Hes' tablets were included there would affect later generations' images of the past and, as in a written account, was a judgment on the relative contributions of the men to the lake.

This episode depicts the importance of form and face in Chinese relationships. The Weis claimed that the Hes' tablets would disgrace the remembrance of Yang and Wei by demeaning their work: such loss of face, they argued, was unconscionable. Many of their technical arguments were based on matters of form, of following the proper procedure. The Hes' supporters argued mainly on the basis of precedence (form provided by the past), never challenging, for example, the procedure of keeping the tablets on a table at the side of the main hall (rather than inside the hall). Wei Qixian's remonstrance in early 1705 that Mao usurped shrine affairs surely lay partially in his own loss of face as shrine keeper. Form, face, and procedure are not simply insubstantial concerns; they provide the essence of the social relationships that exist among the living and between the living and the dead.

In 1710, the Eight Worthies Shrine was built in the clearing to the west of the Shrine for the Virtuous and Kind.[62] Prompted by Lai Jizhi's kinsman, Lai Ersheng, the shrine commemorated four officials (Zhao Shanqi, Gu Zhong, Guo Yuanming, and Zhang Mou) and four county "worthies" (He Shunbin, Fu Xuan [He's son-in-law], He Jing, and Zhang Ling [Sun Zhaowu's opponent in 1519]). We are not told whether the Hes' inclusion in the new shrine meant that sometime after 1705 their tablets had been ousted from the other shrine. The new shrine, however, enhanced the image of the Hes by placing them in the larger context of lake defenders and at a more central place than they were accorded in the former. In a real sense, form became substance, and with the Hes' more integral inclusion, Mao's image of them became the remembered past.

V

VIEW

Sailing on the Lake from Upper Sun Village to Green Mountain, 1783

CHAPTER

Restoring the Lake: Of Mud-Dredgers, Leeches, and Worm Officers, 1758–1809

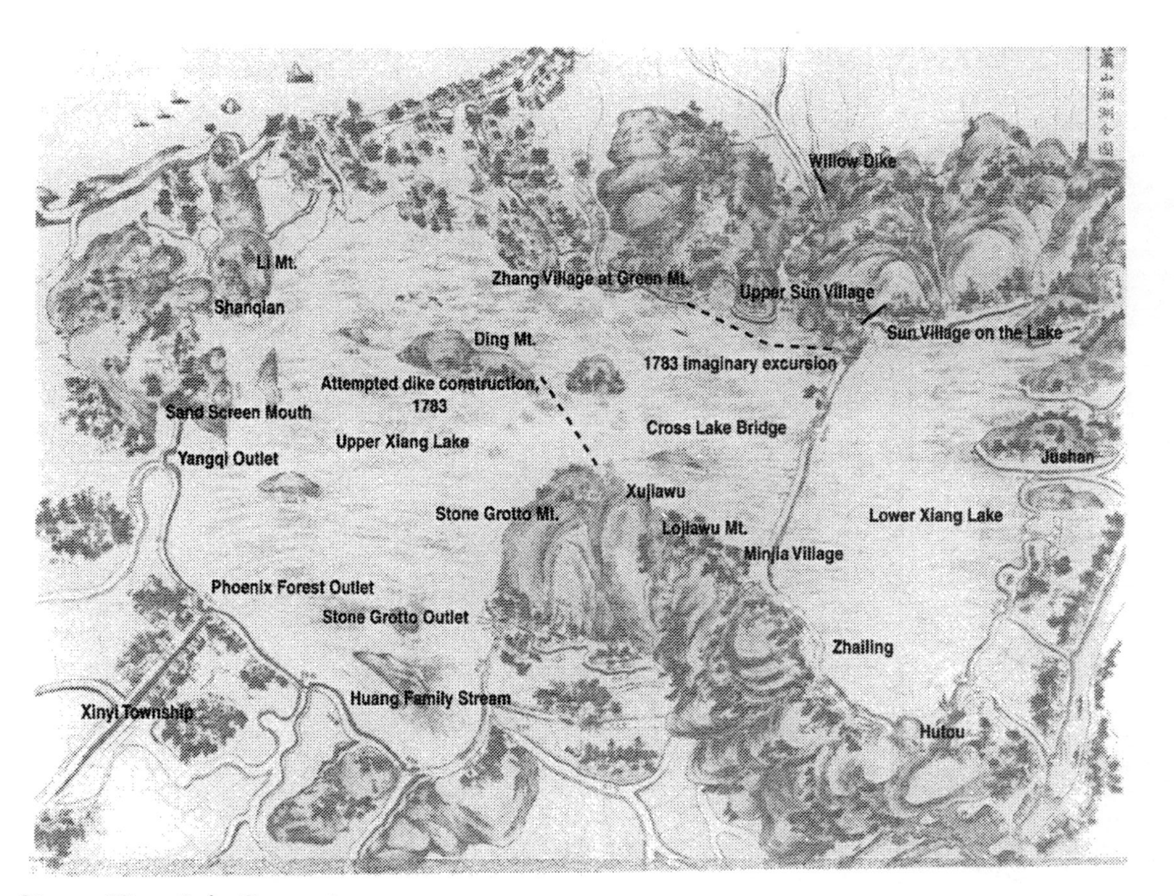

Map 9. Xiang Lake Sites, 1783

VIEW: *Sailing on the Lake from Upper Sun Village to Green Mountain, 1783*

In what was probably the first half of the eighteenth century, Cai Zhongguang's nephew, Cai Weihui, wrote a poem describing a real or imagined boat excursion from Sun Village on Xiang Lake to another site along the western shore. The boat remained close enough to the shore to allow Cai to hear the sounds of music from the mountains.[1] Imagine now a similar trip from Sun Village southward to Green Mountain in the early spring of 1783 as we view local sites from the boat.

Upper Sun Village

Along the shore at Sun Village on the Lake stood the generally cylindrical kilns used to fire the bricks and tiles that had brought initial wealth to the Sun lineage (figure 5). The village was the center of the lake's tile-making industry; twenty-six of the sixty-four lakeside kilns were located here. Sold mostly in Hangzhou and Shaoxing, the bricks and tiles were generally construction materials—tiles for roofs, furnace surfaces, tubing—but the names of the thirty types made around the lake suggest that several may have been more highly finished: Dragon Waist, the General's Tile, Dog's Head Tile, Flowers beside the Dripping Water.[2]

Looking from the lake into the southernmost hamlet, Upper Sun Village, one might see some of the hamlet's eight kilns, named, not very

Figure 5. Xiang Lake with Cross Lake Bridge, Sun Village in background (1920s)

imaginatively, New Shed, Old Shed, Lower Shed, New, Upper Village Big Shed, Lower Village Big Shed, Mid-Valley, and Lai Family.[3] Activity around the kilns would be evident on an early April day, since the industry was situated on the lakeshore (in fact, far inside the original lake border as prescribed in the Song period) and spring was a prime time for firing the kilns. Most villagers were involved in the industry. Mud-dredgers brought the precious resource from the bottom of the lake, and woodcutters supplied not only the fuel for the kilns but also the material for the tile molds. Whether all the Xiang Lake tiles were composed purely of mud or, as in Ducun in the nearby county of Yuyao, were a mixture of mud and decayed reeds is not known.[4] If the latter, then villagers would have been employed as mixers. Others served as firers and carriers both of unfired and finished tiles and of water from the lake to extinguish the kiln fires after twenty-four hours of baking. Because of rainfall, the industry generally avoided operation in the summer; that hiatus allowed agricultural work or attention to more lake-specific activities like fishing.

Although it would be impossible to determine by sight alone the number of households farming or living in what had been lake land, a survey in 1770 showed 109 households in an area of about fifty mu living in houses with tiled roofs or in thatched huts. Also in the area, indications of the longstanding Sun lineage connection to Buddhism, were a Buddhist convent of eight rooms surrounded by a large garden and the Ever Prosperous Shrine of ten rooms.[5]

Mountainside Gravesites

Visible on the mountains to the west and across the lake to the east were many gravesites situated to realize the most auspicious geomantic forces. Daoist fengshui specialists believed that mountains protected burial sites from evil "winds" that blew from behind the graves. In addition, placing a grave on a mountain slope generally provided a propitious, unobstructed prospect in front of the grave. The proximity to water, the source of life, here Xiang Lake, was also a positive feature.[6] Gravesites spread over the mountains, in the words of the lake gazetteer, like "men on a chessboard."[7]

Many gravesites featured large, often omega-shaped embankments at the back and sides of the tombs, conspicuous evidence of the former wealth and prestige of the dead. These masonry embankments, often plastered with white mortar, formed islands as white amid the green foliage as the egret that took the lake as home. Their function: to protect the grave from rainwater that cascaded from higher up the mountain and to keep away any evil wind by diverting it to the right or left of the grave.[8] Such illustrious Lai men as Ming dynasty scholar officials Sanping and Zongdao were entombed on mountains to the lake's west, near their home of Long River, as was Huang Jiugao, Ming dynasty official. Across the lake to the east lay Wei Ji's grave on Pleasant Hill and the grave of Lai Sixing, builder of the Three Goodness Bridge on Snake Mountain.[9]

Cai Weihui's poem had alluded to hearing music from an instrument made of bamboo pipes, but on this early April day the sound of drums and mournful songs might be heard from the mountains. In an essay, Cai noted that such music was unceasing during the days preceding the Festival of the Dead, the only festival whose date was determined by the solar calendar.[10] Falling about April 5, it was preceded by the Cold Food Festival, during which all fires were extinguished for several days. The day before the Festival of the Dead, new fires, symbols of purification and renewal, were lit. On the day of the festival itself, graves were swept, food and incense were offered to the departed, and willow branches were hung on doors or at gravesites to ward off evil spirits.[11]

Not all Chinese were buried so grandly. In the countryside many fields held family tombs, often situated on prime farmland. Many lineages around the lake and in Xiaoshan county established cemeteries for their members. One such cemetery was visible at Zhailing, the home of the Wu lineage: in 1732 two brothers established a twenty-five mu cemetery in memory of their mother, née Lai.[12] Another, established in the late Ming by the Shi lineage, was at one end of the Cross Lake Bridge.[13] People commonly postponed burial until an auspicious gravesite had been found, until a propitious date had been determined, or perhaps until a son or relative had returned for the burial. In two severe floods, in 1770 and 1776, when large sections of the West River dike and the seawall had collapsed, floodwaters carried away many such unburied coffins. The living fared much worse, of course: in the former flood an estimated ten thousand perished.[14]

Green Mountain

The only poem extant about Green Mountain where Zhang Village was located on the lake's west side recounts a visit by the poet Zhang Wenrui to his parents' graves on the mountain.[15] Zhang, an official who throughout his career in the mid-eighteenth century built a reputation as a water control expert, spent his retirement active in repairing the West River dike to protect the county. Zhang's poem, with its images of solitude and desolation, points to the geomantic forces of the gravesite, noting the water cascading from the mountain into Xiang Lake and the not yet withered grass on the slope behind the graves. On Green Mountain lay another famous county grave, that of chaste Mother Hu, née Wo, who reared her children alone after her husband and father-in-law were slain by bandit soldiers at the time of the Manchu conquest.[16] Mother Hu, who remained unmarried, selflessly performing the roles of mother and chaste widow until her death (she lived to be eighty-one), was another significant social role model. Like the recounting of uxorial suicides during the Manchu conquest, her incorporation in the local history upheld the value of female chastity, just as Zhang's poem pointedly perpetuated the model of the filial son.[17]

As with Sun Village on the Lake, a number of households of Zhang Village at Green Mountain were living in former lake land: twenty-one, according to the 1770 survey. Well over half were named Zhang, with the remaining seven households divided among five surnames. One villager, Zhang Zuoming, must have been wealthy, having planted a large tallow tree garden of over half a mu, presumably for commercial purposes. In late autumn the garden would have been striking with the trees' scarlet leaves and berry clusters white as snow.[18] In the garden at the time of the survey lay an unburied coffin waiting for interment.[19] It was a dangerous place to hold a coffin in waiting; in a heavy downpour the former lakeland would be immediately inundated.

Activity near Ding Mountain

On this spring day in 1783, looking from the shore near Green Mountain to the east side of the lake and southeast to the vicinity of Ding Mountain in the middle of the lake, one could see much activity. At first thought, one might assume that the crowds were Buddhist pilgrims making their spring pilgrimages to Hangzhou temples, accompanied by throngs of pleading beggars.[20] But a closer look indicated something ominous. Boats were ferrying back and forth from the eastern shore near Stone Grotto Mountain carrying soil and mud and dumping it into the lake. Already in what was obviously a shallow portion of the lake, the deposited soil had reached the surface. Whether it was noticed by mud-dredgers, mourners, tallow-gatherers, or fishermen, what was happening was obvious: people were building a new dike to connect the shore to Ding Mountain. It would only be worth remarking, as in imagination we disembark at Zhang Village to climb to Tiger Cave up the mountain, that the dike was not conceived at Sun Village on the Lake. Others had embarked on the strategy of dike-building; the history of Xiang Lake had entered a new phase.

CHAPTER: *Restoring the Lake: Of Mud-Dredgers, Leeches, and Worm Officers, 1758–1809*

In many ways the eighteenth century—known generally as the height of traditional Chinese power and splendor—seems a watershed in the history of Xiang Lake. The preceding six centuries acquire the aura of a morality play; men seem almost larger than life: incarnations of vice who plotted maliciously, killed, and seized for self and models of virtue who worked tirelessly, preserved, and saved for others. Xiang Lake became the stage for grand battles between filiality and betrayal, obligation and irresponsibility, loyalty and surrender, chastity and impurity, and civic virtue and private greed. In this light, it is significant that the Shrine for the Eight Worthies, built in 1710, was the last major memorial ever constructed for upholders of the lake and that when it collapsed several decades later, no record mentions that anyone thought it worth reconstructing. A curtain had lowered on the way area people had conceived of the lake in their lives.

In the Song dynasty, though people initially squabbled over water rights, the struggle over the lake had been waged between encroaching officials and defending officials; the people had followed, unwilling or unable to challenge either. From the early Ming, the Wus and especially the Suns seemed larger than life, bringing officials in tow, intimidating elites and commoners alike into abject silence. Mao Qiling's besting of the Suns ended the overt threat of Xiang Lake's becoming Sun Lake. It is not that after Mao no figures dominated the lake's history, nor is it that after

Mao ethical-political issues ceased to emerge. But instead of a morality play, the eighteenth century brought dramas of specific problems with new casts of characters including the leaders of newly emerging lineages and an insatiable supply of sub-bureaucratic clerks and runners. And one theme increasingly dominated the ethical concerns of earlier: the concepts of "public" and "private."

One historian has said that the late eighteenth century with its wealth and power was a period when the "Chinese self-image . . . was . . . as close to the historical reality from which it was abstracted as any such idealized constructs are likely to be."[1] At Xiang Lake it was a period of attempting to restore the lake to its former state, to bring the reality of water management in line with the image of the reservoir as conceived by its creators. In short, the goal was to preserve the lake as a common public good in the face of many small private challenges, which, taken as a whole, endangered it.

Protecting the Lake Area from Nonagricultural Threats

In the decade 1759 to 1768, a number of leading figures from around the lake, led by Huang Yun (jinshi, 1736), instigated action in the name of the public interest to protect the area from private nonagricultural threats. Two—stone-quarrying on Lojiawu Mountain on the lake's east side and the construction of fishing weirs in the Grand Canal—occurred in the larger lake vicinity, whereas the third, the presence of kiln households within former lake land, affected the very body of the lake. All seemed to indicate a substantial entrepreneurial thrust among area inhabitants in the flourishing mid-eighteenth-century economy.

In the last days of 1758 Huang Yun, who had enjoyed a career as magistrate in Sichuan, joined eighteen other men from moderately important lake lineages (Sun, Wu, He, Cai, Zhao, and Wang) in opposing the quarrying of stone on Lojiawu Mountain.[2] Many of these men lived in villages along the lake's east side.[3] One Ding Baoheng had hired some hundred men to begin quarrying at the mountaintop. The opposition's primary objection was that Ding "sought private interests with a black heart"; "not knowing the difference between life and death," Ding's activity "profits one household and harms ten thousand."[4] This pointed stipulation of the clash between the public interest and private greed recalls Mao Qiling's dictum seventy years earlier that private interests (in that case, those of Sun Kaichen) could not serve as the standard for actions around the lake. Though the issue of lake encroachment had always been basically an issue of public versus private interests, Mao Qiling's statement and the language of documents involving the lake struggles of the eighteenth century suggest that by the 1700s the issue was conceived more specifically, or at least was more pointedly expressed.

Ding's opponents claimed specifically that the quarrying would endanger homes and graves on the mountain. Though specific information that these had been or were imminently threatened is lacking, the presumption seemed to be that Ding's activity was moving inexorably toward that end. The opponents also clearly linked Ding's private acquisitiveness with the threat of social unrest: since the quarried stone was of fine quality and quite valuable, the threat of thieves and robbers disturbing the peace was real and upsetting. As with the specificity of the private-public dichotomy, this was the first articulation of the linkage of private actions to the disruption of the whole in social unrest. It would be heard into the twentieth century.

Finally, but importantly, another argument of the restorers resonated with arguments from the late twentieth-century struggle between developers and environmentalists: Ding's quarrying would destroy or damage the county's natural geomantic forces. Memorialists described a curving chain of key mountains moving from south of the Puyang River, past the town of Yiqiao, to Yangqi Mountain on the lake's southern side, to Yewu Mountain in the lake, and finally to Stone Grotto Mountain and Lojiawu Mountain, which sloped to the wall of the county seat. The disturbance of quarrying on Lojiawu Mountain, which was crucial in the geomantic line that ran directly into the seat of county political power, could unleash negative, disruptive forces for the entire county. While fengshui had been used as an argument against disturbing individual gravesites, this is the first reference in Xiang Lake's history to such particular local forces affecting the county or *public* as a whole. It is also interesting that this concern over geomantic balance was expressed at roughly the same time that ecological concepts of equilibrium were first being asserted in the Enlightenment West.[5]

At the base of all the arguments of Ding's opponents lay the Chinese sense of the holistic universe and a marked concern for the public good. The arguments apparently persuaded; Ding and a coquarrier named Yuan were punished and ordered to halt their activity forever.[6]

In summer 1760 Huang Yun addressed another area issue that had become a perennial problem, the placing of bamboo fishing weirs in local streams, especially the Grand Canal.[7] A number had been placed in the Grand Canal and stocked with fish to be raised for commercial purposes. Twice before in the century, in 1715 and 1744, orders had been given to take such enclosures out of streams, though in 1748 their placement had been allowed in some channels. Huang's memorial was based on the holistic interdependent conservancy system. At times of flood, only the Little West River (carrying water to the Three River Sluicegate) and the Grand Canal (emptying into the Qiantang River at Xixing) could drain the county, bounded as it was by dikes and seawall. Effective drainage was essential to prevent the destruction of crops. Because weirs, with their approximately three-inch-wide bamboo slats, were excellent sites for canal-clogging weeds to proliferate, Huang argued that the several dozen weirs in the canal presented a potential danger to effective conservancy. The weirs were ordered removed less than two months after Huang's memorial.

In 1768, continuing his efforts to restore the environment of and around the lake, Huang Yun, joined by unnamed petitioners, asked that the original dividing line between lake and mountain land be restored and that those living inside the lake land be cleared out.[8] The prefect and magistrate, with the help of local leaders, began investigations. The survey of "lake-dwellers," completed in 1770, showed that 308 households were living in the lake land, an increase from the 210 households living there in the early sixteenth century.[9] The inhabitants had increased almost 50 percent over a period of nearly 270 years.

The survey reveals much about aspects of this particular type of insidious encroachment and of society around the lake. Almost three-fourths of the households lived in Lower Xiang Lake, which also contained a similar percentage (72 percent) of the kilns for bricks and tiles. Lower Xiang Lake had been most deleteriously affected by the mid-sixteenth century construction of the Cross Lake Bridge: water moved less easily between the two lake sections, and the lower lake had gradually become clogged with water weeds. In times of serious drought (as in 1689), the shallowest sections might become completely dry. Living between the mountains and the lake on little land, villagers saw the former lake land as additional living space and as an area to which tile industry and paddy land could be extended.

The dimensions of the plots of land suggest that the encroachment came bit by bit; most were only a small fraction of a mu and were often extremely narrow.[10] Wu Wenhua in the village of Hutou near the county seat held a plot of land that was approximately 195 English feet by 5 feet; many similar plot dimensions suggest that as the lake dried up, long but very narrow strips of land next to the shore were taken for private use. In many cases, carrying soil into the area undoubtedly hastened this encroachment. Some plots were larger and suggest households with considerably more wealth. Sun Guoying of Upper Sun Village held two plots (270 feet by 90 feet and 60 feet by 30 feet) totaling over 5 mu of land. At least two households, those of Wu Xianming and Wang Tinggui, held land in the lake beside two villages. Former lake land attracted poor and wealthy alike. While most householders continued to live on the lake's shore, some built homes in the lake. These were thatched huts or more elaborate multiroomed tile-roofed houses; the seven-room house of Sun Beimao and the eight-room house of Wu Weiqi in the lake land off the village of Minjia mark them as wealthy households who both lived in and made their living from what had once been underwater.

Encroaching on the lake was a risky business. If rains came, they might quickly inundate the developed land, destroying crops and flooding homes. When the survey of 1770 was undertaken, for example, plots belonging to villagers in Lojiawu, Xujiawu, and Jushan were underwater and could not be measured. Like the land-hungry poachers on the alluvial deposits of the Qiantang River who often saw their developed land collapse back into the river after a sustained change in currents, the encroachers on the lake were driven by the same desire for a source of livelihood, however risky. Some lake dwellers actually developed land on the top and sides of dikes. Three of the four encroachers from the village of Shanqian on the lake's southwest shore had plots on dikes. All three had planted one or more trees on the land they had seized, a practice far more detrimental to the lake's functioning than simple encroachment: tree roots could begin to break dikes apart, opening the way for serious leakage problems.

The final survey calculations showed that encroachments occurred at seventeen villages and covered between 300 and 350 mu, or approximately fifty acres. The 308 households had built 118 tiled houses, 9 thatched huts, and 5 kilns. Since the original line between mountain and lake land was indistinguishable, investigators used the so-called fish scale register land survey from the Ming to determine whether taxes had been paid on the land and thus whether it had originally been private land or public land (*guan*).[11] None of the landholdings of the 308 households were recorded in the Ming land survey.

While Huang had asked that all encroachers and their property be removed, in effect calling for the absolute exclusion of all private claims on the lakeland, the decision of the local officials (ultimately corroborated by the emperor)[12] carefully balanced public and private interests. The prefect indicated that many of the people had lived all their lives on the lake land and depended on it for their livelihood. If they were suddenly dispossessed, how would they live? And if, on the contrary, the land were recognized as private and the householders were required to pay back taxes, how could they produce the money? Either decision, the officials feared, might lead to social unrest, with households joining together to fight for their interests. The pragmatic and primarily compassionately paternalistic solution was to determine which landholdings were actually blocking the lake's functions and which were on peripheral areas, or "corners," of the lake's shore, producing no real harm to the lake. The decision was to remove between sixty and ninety households and two kilns from the lake land and to instruct village leaders (*zhuangbao*) to maintain stricter enforcement of the lake boundary by reporting any further encroachment.[13] On the emperor's approval, the prohibition against encroachment was carved in stone.

The episode reflects the late eighteenth-century approach of officials to the lake and its inhabitants as pragmatic (preserving the peace by staving off local unrest) and caring (preserving the holdings of 70 to 80 percent of the households). It illustrates the benevolent paternalism of the local officials, who made the magistrate's traditional epithet, the "father-mother" official, a reality. In short, it provides a picture of local government under the empire at its best.

Lai Qijun and the Ding Mountain Dike

An increasingly serious problem at Xiang Lake in the eighteenth century was the corruption and bribery of those charged with local government responsibilities (clerks and runners in the county yamen) and lake surveillance (dike and embankment heads). Though the grand moves of the Song officials and tyrannical lineage leaders were largely absent in the eighteenth century, the lake was now threatened by the massing of countless individual encroachments, often with the connivance of lower level quasi-official "watchdogs" of the public interest. In 1719, for example, lower-level degree holders surnamed Kong and Han from southeast of the lake, bribed county yamen clerks to allow them to pilfer water from the lake at time of drought.[14] Discovery of this pilferage prompted a general investigation that uncovered widespread bribery around the lake: while eighteen official drainage openings existed, villagers had also made thirty-three private openings and more than forty sites for water lifts.

In 1783 similar collusion led to a serious challenge to the integrity of Upper Xiang Lake and to the emergence of another lake hero, Lai Qijun.[15] A jinshi from the Long River lineage, Lai was the grandson of Lai Sixing, the builder of the Three Goodness Bridge. Though that bridge brought few of the harmful effects of the Cross Lake Bridge, Qijun's grandfather's actions had been another expression of arbitrary local power at a time when the lake's continued effectiveness was in serious jeopardy. Qijun's role stands in as much contrast to that of his grandfather as Sun Xuegu's actions in destroying a Guangdong dike stood in relation to Sun Xuesi's construction of the Cross Lake Bridge.

In spring 1783 it came to light that four men had bribed dike administrators Wang Liangqian and Han Shengru not to report their construction of a three-thousand-step earthen dike running from the east shore to Ding Mountain. The four were a current granary clerk named Fu, two dismissed yamen employees, and the holder of a military juren degree. Sources diverge on their goal. The men may have intended the dike primarily to facilitate trade, or, some claimed, they may have intended ultimately to enclose an area to be planted into rice. Incredibly enough, this action was not met by protest; in fact, when Lai Qijun first tried to instigate action against the dike, prominent local figures even provided cover for the guilty. Their reasons are difficult to fathom, given the relatively low social position of three of the encroachers; it is probably significant, however, that the juren holder and dike heads Wang and Han represented three fairly important lineages on the eastern side of the lake. The inertial silence replicates the quietude of earlier lake users whose source of livelihood was threatened. Whether intimidation or simple fear of involvement was at root, the record of the lake shows that area inhabitants were unwilling to act until goaded.

Lai Qijun was made of a different mold than the Song dynasty officials or retiree Wei Ji of the Ming. After his jinshi degree in 1772, he held an official post for only three months; he then returned home for a mourning period and remained there for the rest of his life. He was an active local leader, involving himself in financing a lineage school, in water conservancy, and in flood reconstruction efforts. When he heard about the Ding Mountain Dike, he took a boat out on the lake with his nephew,

Xiangyan, in a pouring rainstorm. Supported by Xiangyan, the barefoot Qijun walked back and forth along the dike, totally exhausting himself but becoming increasingly angry at what was being done and, even more, was being accepted by others. He was so wrought up that he failed to notice the blood running down his legs from the leeches that were sucking on his thighs.[16]

Later that day, calling what was occurring a "deep tragedy," he went to the county seat, where he made an impassioned plea for action to a group of local leaders.

> You give no thought to the lake. For several hundred years we've depended on it for our livelihood. To let it be destroyed in one day is not virtuous (*ren*). The lake's 37,000 mu are to irrigate the 140,000 mu of the nine townships; to rob the lake of 300 mu means 1,400 mu will not get irrigation water. If there is such great advantage and disadvantage so close at hand, not to struggle is stupid.
>
> How can you stop it? If you build a dike, will you not already be reclaiming land; and if you reclaim it, will you not be filling in the water? The lake will silt up and then flood our footpaths, fields, and homes. Gentlemen, how can you continue to be like stone men? We are playing and trifling away time instead of taking charge of the situation. There seems to be no one courageous enough [to act].
>
> Tomorrow you must all accompany me to the magistrate about these matters. If you don't come, then you can't be numbered among this locality's gentry (*shi*). Furthermore, whenever we see you, we'll spit in your [contemptible] face.[17]

The group assented, though there is no indication that Qijun's rhetoric overwhelmed them. The next day, however, with several hundred men in attendance, the magistrate was deeply moved by Qijun's presentation and set out immediately to follow up his charges. He dispatched a yamen runner to arrest the two dike heads, Wang and Han. The two remained silent at first, pretending not to know who was involved in the usurpation. Qijun and others petitioned the magistrate to look in the yamen itself for those involved. Qijun's own determination seemed finally to have ignited the feeling of other local leaders on the issue: sources indicate that dozens wrote accusations and hundreds drafted petitions to see the case through. Finally, as was customary, Wang was tortured to extract a confession, and he provided evidence that the four had bribed him in order to build the dike.

The guilty were not arrested peacefully. On May 15, when the juren holder was seized, many of his supporters gathered. Fights between the dike builders and lake preservers broke out along the road, with some of the latter apparently injured. Qijun, who reached the fight scene near its conclusion, had his clothing torn. But the arrests were eventually completed. The next day the Ding Mountain Dike was leveled. The guilty were ordered to be flogged and put into cangue and fetters. The guilty granary clerk Fu first feigned illness and then tried to use his position to avoid punishment, but the persistence of one of Qijun's kinsmen eventually brought him to justice. Commenting on the conclusion of the case, the magistrate noted that Xiang Lake was no longer one man's lake but the lake of the nine townships, thus stressing the illegitimacy of private over public interest.

Lai Qijun continued to involve himself in local water issues, though his contribution to the lake was concluded. In the severe flood of 1784, he directed efforts to rebuild collapsed river dikes, carrying baskets of soil amid the downpours. The floodwaters eventually became so great that Qijun directed the river and sea dikes to be opened to facilitate drainage. When he fell ill and died a short time later, many suspected that his work on the dikes had exhausted him and brought his early death at age fifty-seven. For his protection of Xiang Lake, Lai Qijun ranks among those whose contribution was honored at earlier shrines. But it was a different age: even his burial site is unrecorded in the county history or the lineage genealogy. His only monument was that, for the present, Upper Xiang Lake was largely free of major encroachment.

Restoring the Lake with Evidential Research: The Work of Yu Shida

Mao Qiling wrote the history of Xiang Lake primarily as a cautionary tale about encroachment on the lake; it praised the preservers and blamed the usurpers. In the 1790s scholar Yu Shida used the writings of Mao and others on the lake as evidence of the lake's original state in order to understand subsequent changes and to restore the lake's effectiveness.[18] The different uses of history and the approaches of Mao and Yu in their work suggest again the seeming shift in intellectual outlook from the late seventeenth to the late eighteenth century. To solve the lake's problems, Yu went beyond the encroachment-ridden past to the earliest recorded details. He approached the lake as historian, surveyor, on-site investigator, sociological interviewer, geographer, and engineer. Engaged in empirical research, he seems immediately a more "modern" man than Mao Qiling and previous investigators.

The wretched state of the lake dikes and the continuing proliferation of illegal private outlets prompted Yu's concern.[19] Both problems had been addressed earlier in the century, the latter, as we have seen, in 1719. Rectifying the state of lake dikes and openings had become the project of Zhao Yuxiang and his son, Zhao Wujin, from 1768 to 1770.[20] Yuxiang was determined to change at least some of the earthen channel outlets to stone sluicegates. Throughout its more than six-century history, earth and stone outlets had alternated in use.[21] Dike outlets were originally closed with earth, but they grew wider from water erosion at each irrigation. More water than allotted particular channels was thereby lost. Outlet dimensions were later set again, and stone sluicegates were constructed. Over the years, these fell into disrepair, began to leak, and contributed to a continual loss of water from the reservoir. They were eventually destroyed, and earthen embankments were again constructed before the openings. Yuxiang believed these earthen embankments to be inadequate in the 1760s. Dikes of earth with earth-filled channel openings were open invitations to farmers to dig their own holes in the dike, over the years leaving those with the job of dike surveillance uncertain of where and how many legal outlets existed. Stone sluicegates would solve the problem of recognizing illegal channel outlets, would ultimately save lake water by ending erosion around the outlets, and would make opening and closing more convenient.

Zhao Yuxiang died before he could complete the installation of eight sluicegates among the lake's original eighteen dike outlets, but Wujin was able to finish the construction. A purchased degree holder, he held planning conferences with other local community leaders to discuss construction methods and the costs of materials and labor. The Zhaos paid for the construction themselves, a total of 2,800 silver taels. The project began in September 1768 and was completed in April 1770. Noteworthy is the prime role played in the renovation of the lake's essential construction by local figures. Just as Lai Qijun almost single-handedly engineered the demolition of the Ding Mountain Dike, so did the Zhaos bring about the overhaul of the dike outlet system. By the late eighteenth century the managerial roles that had been played by active officials and occasionally by retired officials native to Xiaoshan county were being taken by local elites. As in the case of the Zhaos, that meant instigating, planning, carrying out, and even funding efforts at renovation or maintenance.

Local leaders had always been involved with the lake. Dike administrators (*tangzhang*) were assigned to dike outlets to oversee the flow of water, and in the late eighteenth century, three county families had been designated as official overseers of the lake (Xu, Cao, and Zhao).[22] In addition, locals had always been involved in funding or helping with repairs. What differed here was the level of involvement. Even if Zhao Yuxiang and Zhao Wujin were from the "official overseer" family (and that is unclear), their managerial involvement in all aspects of the sluicegate construction transcended the role of overseer. While such control in the public domain by conscientious nonofficial leaders could be seen as the felicitous working of private leadership in the public arena, that same control in the hands of unscrupulous figures might present the possibility of encroachment by a lake manager himself.[23]

A quarter century after the Zhaos' reconstruction efforts, the sluicegates, made of stone slabs and pieces of stone held together with a hardened liquid mortar, began to crack and leak. If twenty-five years is the approximate prerepair life of such a stone sluicegate, the urgency of constant attention to maintenance around the lake is clear. But little attention was apparently paid to repair. A report from one of the Long River Lai lineage pointed to yamen clerks and runners, who controlled the annual repair budget for lake facilities, as responsible for the situation. He claimed that most repairs, if made at all, were inadequate; that eroded dikes were not being buttressed by even as much as an inch of soil; that some eroded dikes were so narrow that people had lost their footing and fallen into the leech-infested waters; and that each day, because of threats of both flood and drought, the situation grew more serious.[24] Yet each year the annual repair fund of two hundred silver taels fell to the grasping yamen underlings, often referred to literally as "worm" or "grub" officers because of their penchant for corruption.

This situation gave rise to Yu Shida's efforts at renovation, beginning with a 1796 petition jointly drafted with Wang Xu, a Xiaoshan lower degree holder who had recently served on the staff (*mufu*) of the provincial governor, Ji Da.[25] The petition followed Yu's trip out to view the lake and dikes (an excursion that "saddened his heart") and a subsequent meeting with Wang.[26] They requested the use of a thousand taels (the equivalent of five years' repair fund) to repair the lake dikes, and they asked that local leaders be given responsibility for organizing the work, managing the money, and overseeing the reconstruction. Claiming that "worm officers" had usurped repair money, Yu and Wang asked official approval of transferring public funds into private hands since those charged with public functions (the yamen underlings) were not performing them. At the end of the five-year local management, control would be returned to official (*guan*) management. Ji Da, who was described as being concerned about the conditions of the people and who was likely moved by a petition cosigned by a former adviser, approved the petition after appointing a deputy to investigate the situation.

With approval (and money) in hand, Yu set out conscientiously to rectify the lake facilities. The process included an extensive investigation of changes that had occurred around the lake, the organization and division of functions into what in essence became managerial committees, and the completion and overseeing of reconstruction. Yu's approach placed him clearly in the important mid-Qing intellectual-philosophical school of evidential or empirical research, which emphasized ascertaining the truth through analyzing sources and revising texts. He based his study on the writings of Gu Zhong of the Song, Fu Xuan of the Ming, and Mao Qiling and Zhang Wenrui of the Qing; he also displayed the evidential school's assiduous attention to epigraphy by collating the large number of former prohibitory carvings and engravings. Wang Congyan, author of the preface to Yu's volume of findings, *A Summary of the Investigation into Xiaoshan County's Xiang Lake (Xiaoshan Xianghu kaolue)*, notes that Yu's approach was painstaking: he attempted to discover the regulations of the past and "to search for the truth in actual facts" (*shishi qiushi*), not content to rely on rumors.[27] Just as the evidential researcher tried to eliminate accretions or distortions in texts that had developed over time, so did Yu use the record of the past to help remove the accumulated impairments on the lake.

Yu's investigation in the field centered on the eighteen original dike outlets for which Zhao Shanqi had devised an irrigation schedule in the 1150s. Since then dike outlets had been dug open, closed, and reopened in many places other than the prescribed eighteen; Yu wanted to learn how the current situation corresponded to the past. His description of investigating the sites of five dike openings not only provides the best evidence of Yu's approach; more importantly, it conveys the extent of changes around the lake and lack of attention paid over the centuries to conditions at the lake.[28]

> *Stone Grotto Outlet.* At the southern foot of Stone Grotto Mountain, this outlet has theoretically been of no use since the

Hongzhi period (1488–1506) of the Ming, when the Maoshan embankment was built near Linpu and the townships of Chonghua, Zhaoming, and Yuhua began to use water from the Little West River.

The local residents are actually opposed to using water from the lake through this outlet. Legend has it that it is the dragon's mouth, and that whenever it is opened, it will inevitably bring a calamity to the people. (Instead, what is done whenever people nearby need water is to drain water illegally from the nearby outlet known as the Huang Family Stream!) There are those who conjecture that because of the "inevitable" calamity visited upon local people by the use of this outlet, the ancient worthies certainly did not construct an opening in this place. Or people even claim (given the area's dependence on the Little West River) that these townships never needed lake water. These are clearly unfounded ideas. We know that Stone Grotto Outlet was established: Was it for nothing?

As it is now, however, there is little use in using Stone Grotto opening, primarily because of the topography of the area and the use of Little West River water . . . .[29]

Willow Dike. Was this a dike or an outlet? There is nothing today called Willow Dike . . . . Gu Zhong, however, recorded that Willow Dike irrigated Xiaxiao township's Fan Harbor Village—then it must have been an outlet. Perhaps it was called Willow Dike because willows were planted near the outlet or because the name came from a poem where the word *outlet* was not included.

In any case, today we find Tangzi Dike in the northwest of the lake, but it was not one of the original eighteen outlets. It does, however, irrigate what Willow Dike irrigated; moreover, Tangzi Dike Village has a Willow Dike temple. Willow Dike and Tangzi Dike must thus be the same.

[To be certain, however,] on August 26, 1798, I went with Wang Xu to visit the irrigation outlet. We rested at the Willow Dike temple, where they were conducting the Buddhist Festival of Hungry Ghosts [the festival for the dead who have no one to care for their graves]. We met local elders who had assembled, and we asked questions about whether Willow Dike was in fact the old name of Tangzi Dike. They said that Tangzi Dike had been built in the late Ming period after the original dike outlet leaked. But, according to written accounts before the late Ming, there was already a place on the lake then called Tangzi Dike. Therefore, the oral tradition of the elders is insufficient proof. However, Tangzi Dike is over ten li from the county seat, whereas the gazetteer places Willow Dike only four li away. This does not add up. We cannot discount the possibility that the written records are also wrong.[30]

Phoenix Forest Outlet. Designed to irrigate over 20,000 mu in the highest township in the county, Xinyi, this crucial outlet must work properly. Irrigation channels here are shallow and narrow; many fields are on hillsides separated from streams. In the irrigation process, it is often necessary to build embankments across streams. Such activity necessarily presents obstacles to traveling merchants and others who use the channels as thoroughfares. Though the embankments are only erected during the irrigation period of little more than a month, merchants bribe dike assistants to open the embankments. Recently there has been a dispute at the county seat over whether embankments really need to be built; if they were not necessary, yamen clerks could not indulge so readily in corrupt practices.

The elevation and topography of Xinyi, however, requires that the Phoenix Forest outlet has the Lian Dike and Tianchang embankment. Because of these embankments, commercial goods have to enter the river dikes at Linpu [several miles to the southeast]; thus licensed firms at Yiqiao suffer great inconvenience.

The yamen clerks and runners have a field day; businesses bribe them to block the building of the embankments or to open them illegally during the period of irrigation. The clerks and runners then charge farmers a "water release fee" to initiate the irrigation. Many times the end of the irrigation period has not been reached when the embankment is simply torn down.[31]

Li Mountain South, Li Mountain North. These two openings used to irrigate more than 1,500 mu in Anyang township. But after the Qiyan excavation project in the 1450s, most of the land of the townships lay south of the river. These openings thus lost their raison d'être. Today there is no trace of them. I do not know, however, when farmers stopped using them.

One other reason for their falling into disuse was the prosperous tile industry here. The presence of the irrigation channels made it inconvenient for carrying mud or tiles. Because of the mud-dredging and the kilns, the lake as irrigation reservoir in this area began to decline. . . .[32]

Sand Screen Mouth. This opening takes its name from its surroundings; it has mountains on two sides that meet like the angle of a folding screen. It is a private drainage outlet only a hundred steps from the lake's Yangqi outlet, yet people from Xu village all consider it an official outlet. They release water to irrigate the fields of Beacon Lake (originally called Qi Family Lake). The small area is actually almost surrounded by mountains; and when the rains come, water pours from them and collects in the lake and is used to irrigate land belonging to those in Qi Family Village and Tingzi Village—both of which were originally lake land but were developed into paddies by people from the mountains.

If there is no rain, people in the area use the private outlet to Xiang Lake or even try to use water from the Yangqi outlet. This must stop. The outlet should be permanently closed.[33]

Yu's research was questioning and empirical. At Willow Dike, he argues with himself over the merits of written sources and the oral tradition, noting the possibility of error in each. He offers a considered judgment on the value of the Stone Grotto opening. Later in his summary he admonishes people on techniques of repairing dikes. He recounts simple experiments he conducted to ascertain the relative cohesive strength and (therefore) value of black mud dredged primarily from around Ding Mountain and the more generally available yellow mud.[34] Yu emerges as a careful researcher, conscious of the many factors and possibilities of reality. Methodical and meticulous, he did not make rapid judgments: even when every source indicated that Willow and Tangzi dikes were identical, he withheld judgment.

More important, these accounts offer more insights into the situation of the lake in the last years of the eighteenth century. They make us aware of the lake in its larger context. At this time, for example, it did not exist for all original nine townships: the Qiyan excavation had sliced off most of Anyang township, leaving the Li Mountain outlets useless. The building of the Maoshan embankment (and Three River Sluicegate) had made outlets on the east of the lake (especially Stone Grotto) unimportant in the context of irrigation water from the Little West River. Yu also noted from his Li Mountain excursion the powerful force and threat of the nearby Qiantang, changing the shoreline, eroding the farmers' very livelihood.

Yu's accounts point to the recurrent "evil practices" that jeopardized the lake, particularly his description of Sand Screen Mouth: cutting private outlets in dikes and encroaching on lake land. In his study of the dikes, Yu found fourteen private outlets; though far fewer than the thirty-three in 1719, Yu called them intolerable.[35] He made it clear that he could understand the desperate farmer's breaking open a dike for a short period to save his dying crops, but he claimed that many people cut openings and left water flowing out of them from the beginning of summer to the end of fall. In the dike repair of 1796 all private outlets were closed, and Yu called local elders together to discuss establishing a system of strict regulation to prevent the bribery of "worm officers" to shelter any offenses.

Yu made more specific recommendations for dealing with the problems of encroachment, which he viewed as particularly insidious: first a willow tree was planted; then soil was built up; eventually even a dwelling was constructed. People planted lotuses and raised fish, refusing to weed and clean out the areas, which eventually became easy to fill with soil. "It is like people who have an illness in the four limbs who contend that their internal organs are not harmed; they do not know if their four limbs are not healed, the internal organs will become worse and worse."[36] To stop the encroachment, Yu argued, the magistrate must order the dike assistants (*tangfu*) and local baojia heads (*dicong*) to meet with local leaders every spring and autumn to inspect lake dikes for private outlets and lake land for rice- or lotus-growing or fish-raising. Such practices should be stopped, the guilty punished, and the dike assistants and baojia heads assigned to the site heavily flogged. Yu was well aware of sociopolitical realities: though authority to end the problems belonged to the magistrate, he said, ultimately the job would fall to the leaders of the community (*jinshen*).[37]

Another lake problem noted by Yu was the deleterious effects of gathering lake grass.[38] Collected during the summer for use as fertilizer, the grass was brought to the shore and piled on lake dikes. It remained there three days in the heat and rotted, its foul-smelling liquid percolating into the top of the dike, which took on the appearance, Yu said, of a beehive. The top layers of dike soil were loosened. To move the rotted grass, the gatherers used iron harrows, which cut into the dike as the grass was dragged along, also dragging away loose soil. This activity harmed the body of the dike, polluted the air with rank odors, and made walking on utilized dikes difficult, if not impossible. The situation was especially bad at Zhang Village at Green Mountain, where many who harvested and prepared lake grass as an occupation simply took over large sections of the dikes. Yu recommended piling the grasses on sandbanks near the shore. Though gatherers would claim it was inconvenient to transport them from there, he argued that in piling the grasses on dikes, they were paying no attention to the public needs. Yu pleaded for strict prohibitions on using the dikes in such a manner.

Finally, Yu's analysis of the lake takes us beyond the reservoir itself to the delivery aspects of the irrigation system. Lakeside dikes and outlets were critical for the regulated storage of water; but once the lake water flowed into the irrigation canals, other facilities—irrigation lifts, sluicegates, and embankments—became crucial. Yu's analysis of the Phoenix Forest Outlet reflects on the proliferation of embankments. Necessary to allow irrigation to proceed in an orderly manner, embankments, like illegal openings in the lake dike, had a way of appearing where they were unnecessary or even harmful to the overall system. Nine major embankments on the channels led from Stone Pond Outlet and East Gate, consecutive outlets on the lake's northeast shore; the channels of the three outlets on the lake's east shore—Stone Grotto Outlet, Huang Family Stream, and Tong Family Pond, none of which was in essential use because of the Little West River irrigation—had twelve major embankments.[39] The two embankments for the Phoenix Forest Outlet were essential for farmers but obstacles to Yiqiao merchants. Merchants assertively opposed these embankments in the late eighteenth century, pointing to the increasing commercial importance of Yiqiao; the merchants are yet another example of the acquisitiveness seen in quarrying on Lojiawu, placing fishing weirs on the Grand Canal, and the tile-making industry.[40]

The Rise of New Lineages

In the project to renovate dikes and dike outlets, Yu and his coplanners formed managerial committees, the first recorded time in the lake's history that local leaders organized themselves in such a fashion. Four separate committees were formed: a committee of the directors of the budget; a committee charged with surveying, planning, and examining account records; and two committees to manage the actual project, grouped according to townships west and southeast of the lake. The sixteen available positions were filled by eleven men. Five were taken by two men from the Lai lineage of Long River and four by the Han lineage of Yiqiao. The Lais dominated the west management committee, whereas the Hans dominated the southeast. Chen Yuchou of the Chen lineage of Lao Lake Village, northeast of the county seat, filled two committee assignments, and the remaining men (surnamed Wang, Ni, Jia, Shi, and Yu) held one committee position each.

If one uses the composition of these committees as a barometer of lineage strength around the lake, it seems clear that the Lais, with their agricultural-commercial base and continuing civil service degree attainment on the county scene, were still paramount around the lake. A Lai was on the budget committee; the two men on the second committee were both Lais; and two of the three on the western management committee were Lais. From 1750 to 1820, the Long River Lai lineage produced six jinshi, seven juren, and fourteen gongsheng.

Although the powerful Suns and Wus, whose might at the lake endured from about 1400 to 1700, eschewed the civil service degree strategy as a vehicle to local strength, it seems that from the late eighteenth century on, civil service degree success and lineage domination in lake affairs were strongly correlated. Appearing increasingly in important lake roles were men from the Han, Ni, and Chen lineages, all of which began to produce large numbers of successful upper degree holders in the late eighteenth and nineteenth centuries. Families in all three seem to have enjoyed substantial commercial success in the last half of the eighteenth century that allowed them to concentrate on the civil service route to solidify and perhaps legitimate existing local economic power. Men from the Han and Ni clearly based their sociopolitical power on commercial success, the Han in Yiqiao and the Ni in New Dike Town between Yiqiao and Linpu. New Dike Town was a center of the salt trade, and the Nis probably participated in that business.[41]

More is known about the Han. As we have seen, the Han had substantial local power during the Song but in the Yuan, Ming, and early Qing were clearly not significant local powerholders. They were apparently able to capitalize on the growing commercial success of Yiqiao in the eighteenth century.[42] Even a cursory glance at the dates when the lineage published new genealogies, a sign, in part, of lineage solidarity, reveals irregular publication until the Qianlong reign and then a new publication for every new reign period.

The Chen was a longtime Xiaoshan lineage, probably since the Yuan dynasty. The lineage had many branches, but the most significant lived at Lao Lake Village at the southern base of Ren Mountain.[43] Originally basing its strength in agricultural pursuits, the Chen lineage, close to the Grand Canal and county seat, also likely shared commercial interests. Its cohesion and wealth were seen in the Chen lineage estate and schools.[44] When at the end of the seventeenth century Sun Kaichen attempted to build a second cross-lake bridge, one of this lineage, Chen Dakui, had the courage to bring accusations against the powerful Suns. Chen leaders had an acute interest in Xiang Lake's effectiveness as a reservoir, because their lineage base depended on lake water for irrigation.

Overseen by the management committees, the work of strengthening the dikes began in May and ended in July 1796, though Yu's study of the dike openings continued into 1798. At completion the dikes stood eight Chinese feet high (about 9.4 English feet) and had base thicknesses of twenty Chinese feet (23.5 feet) and surface thicknesses of ten Chinese feet (11.8 feet).

Dikes and Embankments: Wang Xu's Further Contributions

The cataclysmic unpredictabilities of nature and water seriously damaged the restoration project of Yu Shida within five years. On the Buddhist Festival of Hungry Ghosts in August 1801, the county suffered a huge flood in an immense storm that broke river, sea, and lake dikes.[45] In the aftermath, Wang Xu again petitioned the county magistrate to use 250,000 cash (*wen*) from the annual repair fund for materials and labor to reconstruct the affected dikes. Wang received permission and proceeded to alter the construction of 278 Chinese feet (.62 English miles) of dike using thick slabs of stone strengthened by wood pilings in a pattern known as the "storehouse wall." Wang finished the project in little more than a month, refusing to allow yamen underlings any active role. As with most of the models of those who preserved the lake, sources describe Wang in heroic terms: braving fierce wind and heavy rain and willing to use his own money when funds ran short.[46]

Wang left another important legacy to the Xiang Lake system with a contribution of 5,000 cash to deal with the problem of the Lian Dike embankment in Xinyi township, a serious local issue noted by Yu in his study of the lake. A sluicegate had originally existed in the dike, but by the late eighteenth century it had been destroyed, though its bottom slab of stone still lay in the channel. Each year during the irrigation period farmers had to construct an embankment, action that invariably angered Yiqiao licensed merchants (*yahang*). Plans to construct a new sluicegate had been stillborn in the 1790s due to an absence of funds.

In the irrigation season of 1808, Yiqiao merchants, led by Han Jin, destroyed the embankment; the farmers in eighteen villages then had no irrigation water for their crops. As the plants wilted and died, scorched by the late summer heat, farmers saw their livelihood and their means of paying taxes disappear. Han's arbitrary action sent a wave of unrest through the drought-stricken area. The animosity was transmuted to accusations against Han when Wang Xu singlehandedly argued the case of the farmers to the governor.[47] The governor responded by sending a water conservancy deputy to investigate the site. The deputy's final report to the governor clearly misrepresented the facts of the case, even claiming that there had never been a sluicegate in the dike. "Worm officers" were not necessarily only yamen underlings.

On Wang's new request, the circuit intendant for three prefectures in eastern Zhejiang was appointed to oversee the investigation. The officials decided to build a new sluicegate in the embankment as it had originally existed. The sluicegate would be closed during the irrigation period, during which time merchant boats would be permitted no closer than about one hundred paces away. The gate would be opened for the rest of the year. A stone slab was engraved to this effect and erected at the yamen in the county seat.

The issue, as Wang described it to the governor, was to harmonize different legitimate private and public interests. The embankment with closed sluicegate would continue to be an *inconvenience* for merchants, but without embankment or sluicegate, the future would hold *impoverishment* for the farmers. In balancing the group interests, Wang was guided by the Confucian conception of the state economy: "The country emphasizes agriculture as the root; it would thus be unseemly to consider commerce as the root."[48] He argued ingenuously that merchants would obey fixed prohibitions just as farmers obeyed regulations against private dike outlets to release irrigation water!

After the sluicegate was constructed, its managers brought together more than twenty interested leaders to form the Lian Dike Society (*Lianyan shehui*), a permanent management group to undertake repairs and rebuild the sluicegate if necessary.[49] This organization brought elite-managed water control work around the lake to a new phase. In effect, local leaders were establishing a corporation to manage nongovernmental funds for the interests of Xinyi township irrigators. Before 1809 Yiqiao merchants had found the irrigators easy to overcome; after that year, any such challenges were met by an organized body "connected" by common interests. Another novel and extraordinarily significant aspect of this organization was its attempt to set up funds outside the purview of government administrators or underlings, the worm officers who often kept conservancy funds for themselves. The society members agreed to contribute money to buy land, the proceeds of which would support repair and reconstruction. The property was divided into seven shares with the repair each year undertaken on a rotating basis by each group of shareholders. The society still dealt with the magistrate in seeking the government's help to attain the bottom slabs for the sluicegate. On the whole, however, the degree of local initiative and autonomy in this arrangement points to the crucial role of local elites in water management in the mid-Qing dynasty.

By 1809 the large and small contentions involving the lake raised in the preceding half century had largely been settled. But the issues they raised—the meaning of public and private; the extent of the public interest over and against private interests; determining the relative legitimacy and priority of different private interests; and determining when shared private interests become a public interest—would be constant pressures in Xiang Lake's subsequent history. The issues were not new, but increasingly they were framed in more specific terms.

VI

VIEW

Yewu Mountain, Late 1860s

CHAPTER

The Beginning of the End: Cataclysm and Silence, 1861–1901

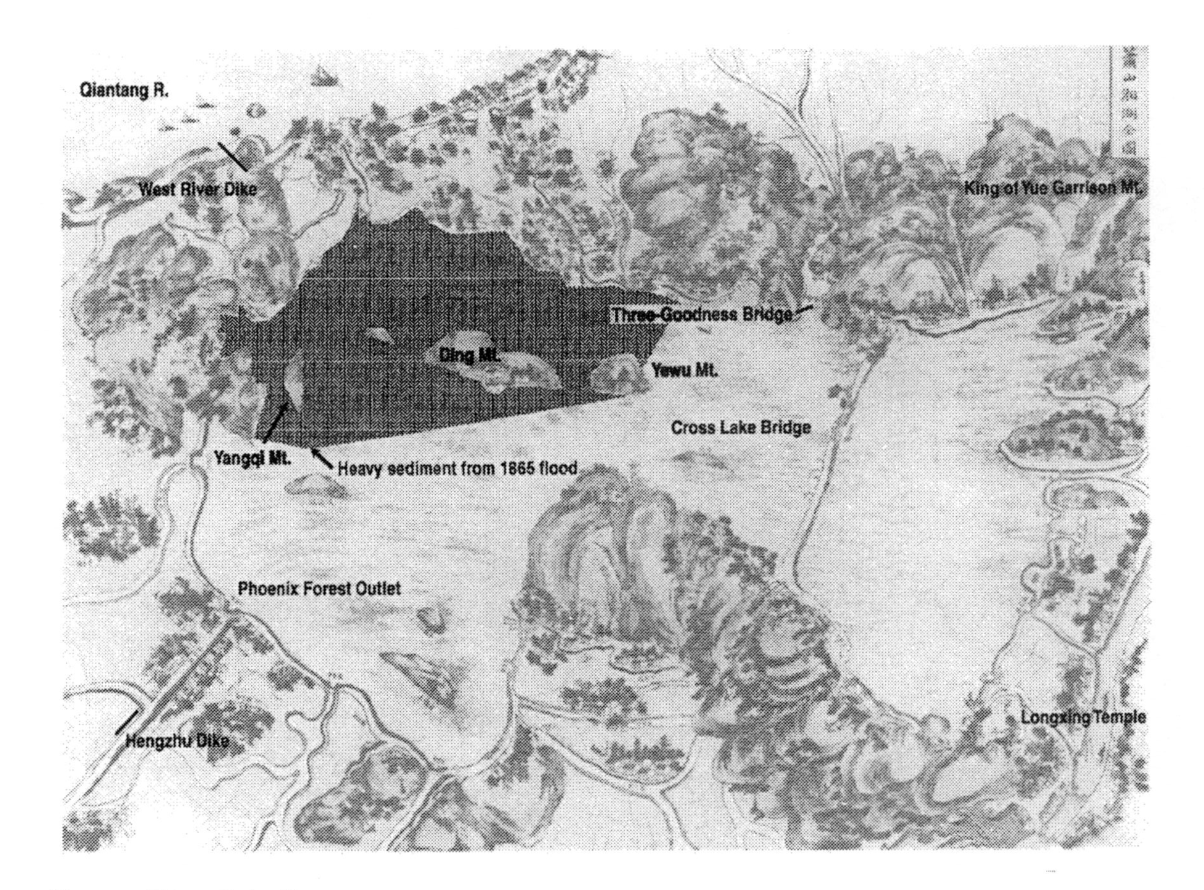

Map 10. Xiang Lake Site, Late 1860s

VIEW: *Yewu Mountain, Late 1860s*

The Mid-Autumn Festival, which falls on the fifteenth day of the eighth lunar month (September's full moon), was one of the year's most merry and light-hearted celebrations. Viewing the bright moon from lake, river, mountain, or family courtyard, eating "moon cakes" just as the black toad (the story went) ate the moon each month, downing great quantities of rice wine, listening to music on the zither or balloon guitar—all were infused with the joy of the coming harvest.[1] It was a night for conviviality.

At first reading, then, feelings expressed in a late nineteenth-century poem by Tang Kanru, "Accompanying a Friend to Yewu Mountain on the Night of the Mid-Autumn Festival," seem markedly inappropriate.[2]

> The cool wind blows the autumn grass
> As the daylight is quickly disappearing in the west.
> While the tops of the distant trees are shrouded in cold mist,
> The setting sun reflects on the thatched roofs.
> Gradually the remaining sunlight is diffused in the distant emptiness.
>
> It is already cold in the valleys.
> The glistening white moon lights the night.
> In the cold, the cicada laments the falling leaves
> As the light wind blows our thin silk clothes.
> Cold dew drips from the autumn chrysanthemum.
>
> We travel to Yewu Mountain.
> A multitude of images dissipate before our eyes—
> Many of the surrounding mountains fade into emptiness.

Reflecting the moon, the lake water shines like dressed silk;
As we move in the water, we sit in melancholy, our hair and beards graying.
When our thoughts turn to the violent changes that have occurred,
Our flowing tears drop like sleet.

People's lives within heaven and earth
Whirl like a floating cloud
As we roam this poverty-stricken autumn....

While many of the ideas Tang expresses—the pathos engendered by the passage of time and the thoughts of native place—are unexceptional in Chinese poetry, what is notable is his linkage of this harvest festival with images of coldness and desolation. The glistening moon seems icy as it shines over chrysanthemums that drip cold dew and men who weep tears "like sleet." The events in Xiaoshan county and the area around Xiang Lake in the early 1860s surely contributed to Tang's desolate vision.

Yewu Mountain, to which Tang and his friend had come, rose at the center of the lake.[3] Of the eight island mountains in Upper Xiang Lake, it was the tallest (over seventy-seven meters); viewing the lake from its summit became a goal of many people on lake excursions. Like all the islands—except for the lowest, named Dirty Island, which pastured farm animals—Yewu beckoned travelers on warm spring days with berries from the arbutus tree, in the heat of summer with its plums, and in the cool of autumn with persimmons and chestnuts.[4] Love seeds, bright red beans that the eighth-century poet Wang Wei had immortalized as a symbol of love, grew on all the islands, attracting the romantic; and the falling autumn leaves perennially attracted the sensitive and poetic.[5]

It is the image of autumn leaves—of melancholy, endings, and death—rather than images of productivity that more closely matches what might be called the historical "spirit" of the mountain. Long before Xiang Lake existed, this mountain had figured in the story of Xiang Yu, the late Qin dynasty warrior whose struggle with the farmer turned military leader Liu Bang was the bloody prelude to the Han dynasty (202 B.C.–A.D. 220).[6] Xiang was so intimate with Fan Zeng, one of his advisers, that he considered him a second father. Fan repeatedly warned Xiang that Liu Bang was the most dangerous of his enemies. Xiang once banqueted Liu after Liu had won a crucial military engagement; Fan tried to incite Xiang to kill Liu while he had the chance, but Xiang refused. Liu later sent gifts of jade discs to Xiang and jade cups to Fan. Xiang accepted the gifts, but "Fan [Z]eng put the jade cups on the ground, drew his sword and smashed them to pieces. 'Bah!' he cried. 'Advice is wasted on a fool. The lord of Pei [Liu Bang] will wrest your empire from you and take us all captive.'"[7]

During later military campaigns, Xiang also spurned Fan's advice to deal expeditiously with enemies. Like Wu Zixu, the ill-fated faithful adviser to the Wu king two and a half centuries earlier, the far-sighted Fan succeeded only in raising the suspicion of Xiang so that he began to take away Fan's authority. Though Fan was angry, he retired quietly; soon he fell ill and died. Local legend had it that before his death, and well aware of the imminent defeat of what might be called his adoptive son, Fan had climbed Yewu Mountain. Overcome by despair and desolation at the severing of his close relationship with Xiang and by the tragic futility of efforts to bring Xiang victory, Fan broke off the southern peak of the mountain and cast it into the Crow River,[8] hence the name of the mountain Yewu, or Thrown [into] the Crow.

The mountain became a symbol of personal desolation, dashed hopes, and imminent defeat. Perhaps as if to point to the Buddhist answer to the vanity and sufferings of human existence raised by this symbol, a Buddhist temple was constructed at the southeastern foot of Yewu Mountain. Although details of the time and circumstances of its construction are lost, we know from poems that it was probably built in the eighteenth century. Poems by Tao Yuancao and Wang Duanlu (son of Wang Congyan, who had been involved in Yu Shida's restoration of the lake) mention the temple and suggest its ties to the Chongfu Yang Temple on Yangqi Mountain (often visited in the seventeenth century by Lai Jizhi).[9] The temple allured both the faithful and those simply on excursions to the island.[10]

The temple was linked symbolically with the lake in two ways. Among the eight prescribed preeminent scenic sights in the county, Xiang Lake's distinctive element of beauty was the reflection of the clouds in its water. The name of the temple—the Clouds of Xiang Temple—connected it directly with this aspect of the lake. In addition, in the Jiaqing period (1796–1821), one Tang Yuanyu erected at the right pillar of the temple a small shrine for all those who had records of accomplishment at Xiang Lake.[11] Though a minor undertaking, this was the only tangible remembrance of the lake preservers established after the Shrine for the Eight Worthies collapsed in the mid-eighteenth century.

In autumn 1861 the Clouds of Xiang Temple was razed along with many other Buddhist, Daoist, and Confucian structures, all destroyed by invaders from the Taiping Tianguo, the Heavenly Kingdom of Great

Peace.[12] The disaster visited on the lake and its environs by these invaders in the years 1861 to 1863 and by a cataclysmic flood in 1865 devastated the people of the area. Yet for us they point beyond these horrifying tragedies to the ultimate doom of Xiang Lake: Yewu Mountain and the burning of Clouds of Xiang Temple presage the final struggle over the lake itself. They are portents of the final defeat as clear as Fan Zeng's own foretelling of Xiang Yu's defeat at the hands of Liu Bang.

CHAPTER: *The Beginning of the End: Cataclysm and Silence, 1861–1901*

There are no indications that the people of Xiang Lake expected the twin disasters of the 1860s. No recent portent had been spotted: an earthquake in April 1853 had not seemed a harbinger of subsequent calamity.[1] Perhaps the Chinese, whether farmer, fisherman, mud-dredger, or merchant, always expected disaster, or at least loss, from the vagaries of weather, the caprice of government functionaries, and the whim of powerful local elites. Although major military attacks in the area had been mercifully few—the fighting of the Mongol-Ming interregnum, the guerrilla raids of the Japanese pirates, and the trauma of the Manchu takeover—banditry was endemic. Robbers with plunder and rape on their minds preyed on the innards of Chinese society just as surely as roaming wolves and even tigers gnawed those of their victims. Spasms of unpredictable violence regularly punctuated lives often already marked by the baleful effects of unscrupulous human predators who took as fair game those outside their circles of family or other "connections" (*guanxi*). For a people living in such a society, continually at risk from nature and other human beings, the concept of "fate" could understandably emerge as an important, if not consoling, explanation of events. It is also understandable that those in superior positions—government officials, the wealthy, fathers, husbands, elder brothers—could self-interestedly use the rationalization of "fate" in dealing with subordinates while stressing those respective virtues proper to their subordinates: submission, respect, filiality, chastity, obedience. Monuments were built for the meek who displayed these virtues and accepted their fates. Late nineteenth-century travelers along the several miles of the Grand Canal immediately north of the lake between Xixing and the county seat marveled at row upon row of granite memorial arches etched with imposing dragons and engraved with flowery inscriptions to filial sons and never remarried (and therefore chaste) widows.[2]

It is nevertheless safe to assume that in the harvest season of 1861 neither superiors nor subordinates foresaw the terror that would change their lives in the next year and a half. In its face the perennial threats to which most had become inured paled in significance; whereas many might be spared the recurrent threats, few escaped the holocaust of human history's most devastating rebellion in its toll of lives and property.

Xiaoshan and the Heavenly Kingdom of Great Peace

Word of the Taiping advance from the south would have reached Xiaoshan before most of the county rice was planted. The rebels took the prefectural capital of Jinhua on May 28, whereas the most common late-ripening rice grown in Xiaoshan was planted in early June. Nothing in the growing season of 1861 indicates that rumors of possible attack in this county so close to the provincial capital altered the planting, weeding, irrigating, and harvesting of the rice crops, the last of which usually began in early to mid-October.[3] Nor is there evidence that the gaiety of the Mid-Autumn Festival in mid-September was dimmed.

The rebellion originated in the early 1850s and had stormed up from the southernmost provinces of Guangdong and Guangxi, fueled by fanatical religious leaders who promised a utopia ("the heavenly kingdom of great peace") to their followers.[4] Shaping elements of Christianity and Chinese tradition into its own idiosyncratic vision, the Taiping movement initially saw success after success in military campaigns against forces of the imperial government. As evidence of their dynastic pretensions, their chief targets were the Manchus, "demons" who had allegedly corrupted the world. They discarded the Manchu-imposed queue—the coiffure of subjection—and grew their hair long. In their march to Nanjing, which they made their capital, many of the tangible structures of traditional society were destroyed as Manchu control was swept away. Once in Nanjing (1853), they forayed to the west, north, and, beginning in the 1860s, to the east. They conquered much territory but never administered it effectively. Their chief legacy was as predator. Just as traditional society defined virtue (submission, obedience, filiality, chastity) to benefit those in authority, so did the Taipings, who built a rigid hierarchy of elites and privileges in the name of brotherhood and equality. Apart from this social predation, moreover, they left a legacy of destruction so severe that a half century later devastated wasteland remained where once-thriving cities and marketplaces had flourished.[5]

The Taiping rebels, like the forces of the Ming founders five hundred years earlier, entered Xiaoshan from the south, thereby avoiding the Qiantang and the surely not-very-efficacious spirit of Li Di; the south, the source of light and warmth, became the path for desolation on the march. They entered Xiaoshan south of the Puyang River on October 27, only a few days after the completion of the harvest. Several hundred people, many probably farmers, led by lower degree holder Lou Bingjian of the most illustrious lineage in Xiaoshan's southern section, died trying to prevent Taiping entry into the county.[6] The Taiping commander made his headquarters outside the county seat at the residence of one of the Chen lineage and began to install the system of Taiping local administration. The main concern was security: guardhouses were established at crucial sites in towns and townships and door registers were issued, naming the legitimate members of each household. The structure of local official administration remained much the same, as did the system of land tenure.[7]

Local Reactions to the Taiping

The reactions of Xiaoshan residents to the Taiping were similar to their ancestors' responses to conquests by outsiders—the Mongols, the Ming, and the Manchus. Hoping to survive the initial invasion, many who were unable to flee or fight hid, some submerged up to their necks amid the rushes along the banks of canals and irrigation channels, shaded by willow, tallow, and camphor laurel. There they endured the attacks of leeches burrowing into their skin—barring some subsequent fatal infection, at least they would be alive. Others fled to the mountains around Xiang Lake or farther to the south where rough terrain might dissuade Taiping pursuit and where a secluded ravine might provide a temporary haven.[8] Some of the wealthy escaped to more secure refuges. Jinshi Han Qin of the now-powerful Yiqiao merchant lineage took his aged mother on a ship to Guangzhou; Chen Yixian (whose nickname was Xiang Fish) escaped to Shanghai after the Taiping murdered his mother, wife, and children.[9]

The vast majority who could not or did not flee had a range of responses. Some, as in earlier crises, killed themselves. Lai Qijian, a juren degree holder, committed suicide in an expression of filiality: he had not been with his aged father when their home was plundered.[10] Many chose self-imposed death by water over Taiping fire and sword. The lake again became a tomb for women who feared ravishment by the long-haired rebels.[11] Cai Shaonan and his wife drowned themselves in a well.[12] The biographies of many in the local gazetteer indicate that before they were killed or injured by Taiping forces they screamed loud curses of defiance, the only resistance they could summon.[13]

Many did muster resistance against the barbaric Taiping invaders. In the absence of effective imperial government action, local defense became a matter for lineages, who formed their own militia units (*tuanlian*) to protect their home areas.[14] Shi Fulan, a nineteen-year-old purchased degree holder, led units from his lineage to defend the county seat; none from his lineage survived.[15] He Fuming, scholar and education officer, and Tang Xueju, relative of the famous early nineteenth-century official Tang Jinzhao, formed lineage militia units to protect the town of Qianqing, southeast of the county seat; both were killed.[16] After the old and the weak were evacuated from Chen Village on the Lake, juren Chen Yirui commanded militia forces of five hundred in a stand to protect their village. Although many Taiping braves were reported killed, every defender died.[17] Yu Bozhi and the thirty men in his militia unit were all killed in two battles near the county seat.[18] Lai Jinfang managed a large force of his lineage and "won" an engagement against the Taiping when the rebels became befuddled by a dense fog and fled, believing the militia stronger than it was.[19]

The early months of 1862 saw bitter fighting and heavy loss of life in Xiaoshan when so-called righteous soldiers in militia units tried to retake the county seat and ports along the Puyang: Linpu, Yiqiao, and Wen Family Dike. As the Taiping brought in reinforcements, corpses of righteous soldiers and the civilian population piled up. Worse yet, military resistance succeeded mainly in enraging the Taiping occupiers, who took revenge by torching each village in the area, increasing the monumental death toll.[20]

With this tragic record of defeat and escalating casualties, many in the

Xiang Lake area understandably turned to an initially promising resistance effort in Bao Village in neighboring Zhuji county. Led by one Bao Lisheng, a twenty-four-year-old villager with a reputation for supernatural powers, the Bao Village resistance drew widespread support from areas whose militia efforts had proved fruitless against the Taiping.[21] Men flocked to Bao Village: the Chen from northeast of the lake; the Huang and Jiang from southeast of the lake; the Hua from near the Puyang River and the Lou from south of the river; the Qu from near the town of Linpu.[22] Several thousand men bearing hoes and shovels traveled to Bao Village via two routes; one force was led by Huang Zhongyao. Huang, a descendant of the sixteenth-century water conservancy expert Huang Jiugao, was a merchant with a penchant for military strategy. In Hangzhou in 1860 he had met a general through whom he became involved in provisioning military units. In the initial Taiping assault on the county, his mother was killed and two of his brothers had been taken prisoner.[23]

In spite of the desperate hopes stirred by the possibility that Bao Village would turn the tide, what occurred there was a disaster. In April or May 1862 thousands of Taiping soldiers descended on the village, which was surrounded by a bamboo palisade, and laid siege during the hot summer.[24] Many died as supplies of food and water ran out. Overcome by powerful thirst, those under siege were reduced to drinking blood sold by the half cup. Epidemic spread in the summer's heat. Finally on August 6 the Taiping broke into the village, slaughtering many who had escaped starvation and disease. Most from the Xiaoshan militia units were killed. While this effort stands in the tradition of "righteous" loyalty and obligation, the endeavor was remarkably ineffective, and, much worse, almost all units were destroyed. Huang Zhongyao escaped Bao Village to form new forces to battle the Taiping near his home village, but his escape seems to be the exception.

Liberation from the Taiping scourge came from the east when joint Chinese and foreign (French and English) forces pushed toward Shaoxing from Ningbo. Shaoxing was recaptured in mid-March 1863. Xiaoshan was retaken by government soldiers shortly thereafter. Militia units continued to participate in attacks on the Taiping as they retreated. He Chuanceng led a unit in a bloody skirmish on the shore of Xiang Lake at the Cross Lake Bridge; in keeping with the general course of most such units, He and his soldiers were all killed.[25] In spite of the Taiping's expulsion from Xiaoshan, their continuing control of Hangzhou (until March 31, 1864) meant the necessity of continuous vigilant defense for county river ports.

The Taiping Legacy

Population statistics reveal the staggering loss of life: in 1850 thirty million people lived in Zhejiang province, but only about eighteen million lived there in 1873. In areas directly affected by the rebellion, population loss was estimated at 40 to 85 percent.[26]

Statistics hide the human face of the tragedy. Deaths recorded in genealogies and county history point to the heartlessness of the bearers of brotherhood and indicate that suffering at their hands was not limited to one social class. Civil service degree holders, community leaders, defenseless widows, the sick and elderly, the wretched poor—all became victims.[27] In many cases, the recorded deaths were ennobled by efforts to uphold the virtues of filiality and female chastity. One Zhang Fengfei pleaded to be substituted for his father, who had been taken captive by the Taiping. The captors responded by killing both the thirty-five-year-old Zhang and his wife.[28] The wife of one poor man living at the house of a Mr. Li, Xu Gao by name, was raped by two Taiping braves. Xu and Li caught the rapists and killed them. But the Taiping leader had Xu and Li arrested and killed and ordered that their heads be hung in the village as a warning.[29]

While local histories recorded the deaths of some who displayed the model virtues, the lives of countless men, women, and children were snuffed out with little notice. In addition to those killed in the sanguinary fighting, people were at continual risk from "rebel deserters and disbanded braves" (former militia men) straggling through the countryside looking for wealth and women.[30] Female infanticide, always linked markedly to economic distress amid natural disaster or civil war, was rampant during this time.[31]

While the vast majority who died were commoners, the rebellion resulted in the death of many of the area's brightest and most able leaders and potential leaders. The extent of this loss becomes evident when one examines the number of men from thirteen of the county's strongest lineages who were killed in the militia fighting.[32] In the thirteen lineages, about 30 percent of all the highest degree holders (jinshi and juren) were killed. While the percentage of degree and rank holders in Chinese society is estimated about 2 percent of the population, the number of degreed and ranked elites recorded killed in these lineages was about 34 percent. This indicates the greater likelihood of local leaders' involvement in the effort to defeat the Taiping threat. The county history indicates that for all lineages the greatest casualties were suffered among those who lived in the county seat area (Cai, Chen, and Wang) and along the Puyang River (Han, Huang, Jiang, Kong, Ni, and Zhou).[33]

The rebellion exacted human, if not mortal, costs in other ways. An episode in the life of Wang Yueyan (1851–1916), who eventually became a wealthy merchant in Xixing, reveals the danger of social predation in a new social order.[34] In the Taiping autumn of 1861, when he was only ten, his family had fled Xixing to hide at Long River, west of Xiang Lake. When the invasion was over, the family returned to Xixing, where they were accosted by one Lu, who had a private grudge against the Wangs. Lu had become a township leader under the Taiping and proceeded to use his new authority to arrest Wang's father. Wang reportedly tried to maneuver another local man named Dai, who was also now connected to the Taiping, to intervene for his father, but to no avail. Wang's father remained a prisoner until the Taiping defeat. In any society experiencing great social and political change, such episodes may be unexceptional; but they were even more likely in a society where connections to power holders was the underlying determinant of attaining one's goals.

The immense human loss of these years was compounded by the physical devastation of the region. One observer graphically described the scene. The years of the rebellion

> had even altered the face of the country; destroyed its communications; deflected its rivers; broken down its sea defenses. During its continuance smiling fields were turned into desolate wildernesses; "fenced cities into ruinous heaps." The plains . . . were strewn with human skeletons; their rivers polluted with floating carcasses; wild beasts descending from their fastnesses in the mountains roamed at large over the land, and made their dens in the ruins of deserted towns.[35]

In many areas destruction came by fire. When they invaded the county, the Taiping burned towns and villages along the Puyang riverbank.[36] Lineage property, including ancestral temples of the Chen, Han, Wu, and Jiang lineages, was burned to the ground.[37] In the county the Taiping torched the county yamen with all its offices, police offices at Fish Lake, Xixing, and Hezhuang, the salt office, and no fewer than thirty-seven Confucian, Buddhist, and Daoist temples and shrines.[38] At the lake itself fire destroyed the Longxing Temple at Pure Land Mountain, a temple on the King of Yue Garrison Mountain, the Chongfu Yang Temple on Yangqi

Mountain, the Clouds of Xiang Temple, and three Buddhist shrines on the lake's southern and western shores.[39]

The continuing livelihood of the lake area, however, was dealt its most serious blow by the destruction of lake dikes and sluicegates, all of which were severely damaged. The Taiping's apparent joyous destruction of lake facilities was certainly facilitated by the negligent maintenance of the preceding seventy years.[40] Not since the work of Yu Shida and Wang Xu had major renovation been undertaken. The Lian Dike sluicegate, part of the Phoenix Forest Outlet system in Xinyi Township, completed through the aggressive leadership of Wang Xu in 1809, was destroyed.[41] In a fierce display of fanatical vandalism, the sluicegate's frame and base were torn out. All the Lian Dike Society's records from the previous half century were destroyed. Not until 1864 were the surviving members of the society able to meet to discuss repairing the facility. For those years, with the lake unusable, irrigators depended completely on rainfall for their crops. For Xiang Lake, the Taiping holocaust was apocalyptic, the beginning of the lake's end as it had existed.

In a poem Han Qin, whose wealth had enabled him and his mother to escape the Taiping horrors by sailing to Guangzhou, used the water-shield plant, one of the lake's most famous products, as a symbol of Xiang Lake's plight. The setting was spring, when the water-shield was at its most abundant. On the third day of the third lunar month, according to tradition, farmers used the sounds of the water frog to foretell the harvest's abundance. These images juxtaposed to the lack of prosperity in the wake of the Taiping lend the poem considerable irony.[42] The date of Han's excursion: April 18, 1874.

Floating on Xiang Lake on the Third of the Third Lunar Month

A sparkling dressing case, like clear shimmering jade in the full glare of the sun.
We drift with the partridge, following the painted sail.
The rising white water floats into shore—
There, the myriads of flowers are red beside the mountain village.
Now we hear the sound of the water frog, a prophecy of the harvest.
We have a feast of fish; I urge my guests to eat and drink.
We talk about the water-shield plant, silken and delicious:
But of thirty years' spring dream, there is no trace.

Han provides a comment:

In 1848 in the third lunar month, I accompanied Fu Zhitang to the King of Yue Garrison to cook some water-shield plants, drink wine, and take in the view. Now, twenty-six years later, the temple at the garrison has been destroyed by the Taiping. The water-shield plants in the lake have been pulled up by the roots to be devoured by the starving people. The lake, a decade after the victory over the Taiping, still produces very little.

The Floods

According to the holistic unity of the worlds of nature and humanity assumed by the Chinese, nature itself would reflect the Taiping cataclysm, for such a human-produced calamity meant that in the universe as a whole something was terribly awry. The desolation and suffering of the first winter after the Taiping seizure of the Xiang Lake area was indeed heightened by nature's own expression of desolation: a tremendous blizzard reportedly left snow six feet deep on level ground, and bitter temperatures froze the lake and even Little West River.[43]

Far more serious in the long term, however, were three floods that raged, like the Taipings, out of the mountains from the south, each more devastating than its predecessor. In the afternoon of the day before the Mid-Autumn Festival in 1850 (September 19), with people in the middle of preparations or, at least, thoughts of the holiday, the West River Dike, which held the Qiantang and Puyang rivers, broke.[44] The floodwaters, which came just as the rice was beginning to ripen, inundated the fields and destroyed much property. Under the leadership of Wang Lianxi, long active in water control management, the large Dragon's Mouth Sluicegate at Xixing was opened to allow the floodwaters within the dikes to drain.[45] There was no harvest that year; the price of rice rose rapidly as the number of hungry multiplied. On June 30, 1862, during the Taiping occupation of the county, the West River dike broke once again, with reports that water in many places reached a depth of almost six feet. There is no evidence that repairs were made, though when breaches had occurred in the 1830s and 1840s, measures were taken to reconstruct affected dike sections. Taiping control and the sporadic fighting it engendered focused attention on daily survival, leaving little time for thoughts of major dike reconstruction.

The most devastating flood struck in June 1865. That month the steady rains that obscured the outline of the mountains to the south were punctuated by frequent heavy nighttime thunderstorms whose menacing lightning reflected in fields and lake alike. After more than two weeks of rain, the river tore through the West River Dike in thirty-seven places to the south and southwest of Xiang Lake.[46] The torrents of water swept away all in their path: houses, temples, even the Three Goodness Bridge in Xiang Lake.[47] Taiping-damaged sluicegates disappeared completely in the tide as dikes, already weakened and in disrepair, were torn open by the rushing water. People used small boats or even wooden tubs, frantically trying to reach higher ground. The cemeteries high on mountains around the lake attracted many of the living to find safety among the dead.[48] Symbolic of their import, the floodwaters carried away the bloated corpses of the drowned and the coffins of the dead that awaited burial.[49]

For the history of Xiang Lake, even more significant than the damage caused by floodwaters was the load of silt they deposited around Ding Mountain in the middle of Upper Xiang Lake. The sediment covered over five thousand mu (over eight hundred acres) of lake land from Ding Mountain northwest to Zhang Village at Green Mountain and south to near Yangqi Mountain. When water in the lake was abundant, the sediment remained submerged, but the alluvial deposits became visible if the water level was down. This situation was especially alluring to ambitious men with dreams of land reclamation.[50] By the harvest season of 1865, then, nature had put its imprimatur on the Taiping destruction of the lake; it had set the stage for the final struggle over the reservoir.

Reconstruction Efforts on West River Dike

The river dike formed the focus of the reconstruction efforts undertaken in the wake of the rebellion and floods. So long as major repairs went unfinished, people's lives and property, as well as the lake, were in constant jeopardy and the lake became of secondary importance. At risk of flooding with the broken West River Dike were not only the plains of Xiaoshan but also those of Shanyin and Guiji counties, all the territory within the dikes of the Three River drainage system. Work on river and sea dikes in the counties (Shaoxing prefecture) had traditionally been funded by local taxes or contributions, and after 1586 such projects had been managed by local leaders.[51]

In spite of their natural hydrological linkage and common interests elites from the counties often seemed unwilling to share in the expenses of what they considered mainly the responsibility of others. Shanyin and Guiji counties treated the West River Dike, which bordered only Xiaoshan, as the latter's primary responsibility, whereas Xiaoshan treated the Three River Sluicegate as Shanyin's and Guiji's primary responsibility. In

1834–1835 leaders from the two latter counties initially refused to contribute to the repair of a serious break in the dike. After considerable debate and stalling, they agreed to contribute sixty wen per mu of paddy land to Xiaoshan's seventy-five wen per mu.[52]

In 1865 Duan Guangqing, an official who had been at work on water- and alluvial-related issues in the area, set out to organize the dike repair project; he faced immediate difficulties with funding and management.[53] Elites from the three counties immediately fell into the same natural factions on funding the repairs. Those from Shanyin and Guiji wanted to divide the cost equally among the three counties rather than to split it according to irrigated mu. Such a plan would have saved the two counties the equivalent of a tax on about sixty thousand mu.[54] They eventually decided on the old method of figuring tax per mu. Even with the assessed funds, local leaders had to contribute substantial sums of their own.

Duan's other major concern was management. He was well aware of the elites' own day-to-day difficulties even in matters of livelihood in the wake of the traumas of the past half decade. If public facilities had to be renovated, many people's private affairs and losses also had to be dealt with. Duan found elites querulous and self-interestedly evasive, but he knew that management in the hands of yamen underlings would be disastrous. From the government's viewpoint, local management would also provide a likely financial cushion for the project: if tax funds ran dry before the project were completed, a local figure with an interest and much time already invested in the effort would probably contribute sums for either material or labor. In the end, Duan convinced the local leaders that they had no alternative to managing the projects. Duan himself agreed to oversee the reconstruction. The work began at a large break at Wen Family Dike, near Upper Xiang Lake's southwestern shore. Elite managers gathered workers, purchased timber, and divided the work into seven or eight sections, to each of which a deputy (paid by the provincial government) and local managers were attached.[55]

Reconstruction Efforts at Xiang Lake

After the careful restoration efforts of Yu Shida and Wang Xu in the last years of the eighteenth century, Xiang Lake had entered a long period of general neglect. Yu and Wang had arranged control of the annual repair fund for two five-year periods, but after their successful reconstruction, management of the fund reverted back to yamen underlings. By the mid-nineteenth century, the sources are clear that the fund existed in name only. County clerks simply used the money for their own purposes.[56] The dike "elders" (*tangfu*) in charge of dikes and sluicegates, moreover, were interested primarily in enhancing their wealth and accepted bribes to make illegal dike openings, to release water at improper times, and to allow encroachment on dikes, on embankments, and in the lake. Even before the floods made large alluvial deposits, the lake had decreased in size. Lake dikes had become narrow, eroded by people, animals, weather, and rotting water grasses and the harrows that pulled them. They were described as dangerous; in the rain, people often lost their footing and fell into the lake.

In autumn 1844 jinshi Zhang Baikui, joined by a number of unnamed petitioners, asked, using the precedent of Yu and Wang in 1796, to borrow and manage several years' repair funds to restore the dikes and remedy the obvious lake problems. They received no response to the petition. Through the centuries local officials had been the staunchest protectors of the lake, but by the 1840s they were paying little attention to the irrigators' needs. The dikes and sluicegates that collapsed following the Mid-Autumn Festival flood of 1850 precipitated new petitions in 1851 and 1852 for repair of the "dangerous places" along the dikes by making use of the Qiantang River dike collateral fund. Nothing was done. In 1863 and 1864, following the Taiping debacle, magistrate Dai Mei met with local leaders, mostly of the Lai lineage, to discuss lake matters; but the focus was encroachment on the lake and the corruption of sub-bureaucrats and dike chiefs. No action resulted other than another carved inscription prohibiting encroachment.[57] Astonishingly, the calamitous flood of 1865 brought no change. Clerks and runners spent the repair funds, but "not one inch of soil was added" to any of the dikes.[58] In the succeeding fifteen years, whole sections of lake dikes were demolished. Lake water drained out: the system was largely destroyed. When paddies needed irrigation water there was little, if any, and when heavy rains came, no dikes prevented floods.

The cycle Gu Zhong discovered in the 1180s—of alternating calamities of flood and drought—reappeared. The year 1871 saw a big flood. In summer 1876 a scorching drought dried up the rivers. As rice plants drooped and turned yellow, there were reports of relays of men digging frantically around the clock for water in the still-wet middle channels of rivers and creeks. The heat and drought seemed also to have affected people's thoughts, giving rise to all sorts of rumors of unnatural happenings. Most upsetting was the report of demonic paper people, "cut out by unseen hands with magical art," who were entering on the area from the northwest, cutting people's queues (a queueless person was fated to a rapid death) or crushing people to death as they fell from the sky with the force of lead.[59] People stayed awake in groups through the sultry summer nights beating gongs to keep away the fierce paper people. Three years after this threat passed, yet another flood struck.[60]

As even more of the remaining dikes disintegrated, in 1880 five local leaders—two Lais, two Wangs, and a jiansheng-degreed Yu—petitioned the magistrate for action on the lake dikes, once again calling attention to the control of the repair fund by county yamen clerks and runners and the lack of any effective large-scale repair. Apparently some minimal repairs had been undertaken, but not enough to rectify the situation. The reception of the petition might euphemistically be called bureaucratic deferral. Magistrate and prefect embarked on a lengthy period of investigation, of hand-wringing over lack of funds, of dispatching on-site dike visitors who downplayed the severity of the problems, and of exhorting local leaders to assist in handling the situation. In the end, the assistant magistrate was appointed to oversee the repair; but there is no record that repairs were ever undertaken. At a time when funds and effective methods of management were unavailable and when corruption was rampant, bureaucratic stalling became the officials' retreat.

In spring 1887 the West River Dike broke in three places; water rushed into Xiang Lake, collapsing embankments, dikes, and sluicegates. Nothing was done for a year. In spring 1888 a new magistrate, Song Zhiceng, took office and saw immediately the urgency of significant repairs. He appointed distinguished retired county figure Wang Kunhou to manage the repair of three dikes with a total length of over one mile, increasing dike heights and thickness. Funded from the river dike fund, the annual repair fund, and a tax of two wen per mu on land getting water from the lake, the work was completed in the spring of 1889.[61] The episode points once again to the crucial role of the official in lake reconstruction. Although Song's work was relatively minor in the context of the major lake problems, he was able to accomplish some essential repairs. He did so by relying on a retired official in the mold of a Wei Ji or a Mao Qiling, not on lineage elite managers.

Why, we may ask, were local leaders who had been deeply involved in managerial situations since the mid-eighteenth century so quiet in the last third of the nineteenth century? Why did they not assume responsibility for reconstruction? Leaders of key lineages had been the main actors around the lake. They operated to enhance their interests and those of their lineages. A necessary connection never existed between such local leaders and the public interest in traditional China, only contingent connections. In the eighteenth century these elites had acted to champion the perceived public interest of perpetuating the lake. In the comparative wealth and stability of the time, the leaders of various lineages (Huang, Zhao, Lai, and Wang) seemed to act disinterestedly, vigorously relying on an ethos that kept alive the notion of the public interest and public service.

In the late nineteenth century, however, the military and natural cataclysms brought crushing economic and personal hardship with substantial demoralization. Though local elites might return to managerial roles in welfare and education, such actions required far less economic or managerial wherewithal than dredging a large area of a lake and rebuilding its dikes and sluicegates.[62] In the arena of water control, then, there was no one to articulate the public interest; unless an assertive official like Duan or Song emerged, no action was taken.

The Hengzhu Dike Case, 1901

Aggressive local leaders might emerge to set forth an agenda of lake activity with the sole purpose of enhancing personal goals. In 1901 a father and son of the Huang lineage moved to extend their considerable power in an episode involving the Hengzhu Dike southeast of Xiang Lake. When a new magistrate, Qu Zhuo, took office, he was met almost immediately by a water-related proposal from juren degree holder Huang Yuanshou, the son of Huang Zhongyao, a businessman who had led groups of anti-Taiping militia units to Bao Village in 1862. Huang the father had spent a distinguished career as a manager in local affairs.[63] He was involved in the charitable burial of unburied coffins that had washed away in the flood of 1865. He managed repairs on the West River Dike, established a lifeboat system at Wen Family Dike, and built granaries. For many years he was one of two dike administrators for Xiang Lake. The proposal of Huang the son dealt with one of the lake's collateral dikes, the Hengzhu Dike, off the Phoenix Forest Outlet. Huang proposed its destruction and replacement with a bridge and sluicegate.[64]

In Huang's view, the dike presented an obstacle to commercial convenience. Merchants could enter the dike-enclosed area of the county through the sluicegate in the Lian Dike (which had been repaired after its destruction by the Taiping). But with the Hengzhu Dike in place, merchant boats had to travel about sixty li (twenty miles) rather than twenty li (less than seven miles) to the county seat. The new sluicegate would be opened and closed simultaneously with the Lian Dike sluicegate, and Huang contended that no one would be harmed by his proposal, which he had broached with the previous magistrate. Before his transfer, that official had approved joint discussions among the area's managerial elite (*shendong*), led by none other than Huang the father. The petition was cosigned by a number of area degree holders surnamed Huang, Han, and Yu.

Magistrate Qu quickly made clear his independence of strong local power holders. When memorials opposing Huang's proposal reached him from local leaders all surnamed Han, Qu set out to investigate. The Han memorial claimed that the project would harm the livelihood of the area's inhabitants. Qu's investigation corroborated that accusation. He specified in his rejection of Huang's request that such a project would destroy the homes of people along the river channel because the river would be widened to build a bridge under which boats could pass; it would destroy some people's ability to pay taxes because their land would disappear; it would harm the geomancy of the county, bringing negative implications for people's fates; and it would inevitably result in the loss of irrigation water from the lake for Xinyi Township.

He further accused Huang of two errors. The first: falsifying history. In his petition to the new magistrate (whom Huang perhaps felt he could hoodwink), Huang with great impudence miswrote the name of the dike as Huangzhu Dike, arguing that his ancestor Huang Jiugao had built it in the Ming dynasty. For an unsuspecting and less astute new magistrate such an assertion would seem to give Huang's proposal the stamp of legitimacy. Qu showed, however, that the dike had existed since the Song period and that the Huang lineage had nothing to do with it. More serious, Huang had claimed that dismantling the dike would have no effect in the current water level of the channel on either side of the dike, that no difference existed between the upstream and downstream levels. Qu's investigation showed an eight-inch difference in the water level on the sides of the dike. If a bridge replaced this dike, which was critical in regulating the level of water, the fields of thirty-six villages would be harmed.

Huang was so presumptuous of getting his way that he had given orders to dismantle the dike under cover of darkness. Qu's discovery of this action led him to order an immediate cessation. Huang was furious, and he petitioned with his father, criticizing the magistrate's findings. Qu found himself inundated with more petitions, some from those who had cosigned the original with Huang and some insulting ones from Chen lineage leaders (from near the county seat) who patronizingly suggested that the magistrate was new to his position and thus did not really understand the situation. For Qu, the issue was Huang's attempt to aggrandize his private position at the expense of the public in the thirty-six villages. Huang's goal was the same as Sun Xuesi's in the 1550s: to make possible a more convenient thoroughfare to enhance his business—the Huang lineage home was on the route that would benefit from the change.

It later came to light that the Huangs, father and son, were involved in more than the earlier described deceptions, because some of the original cosigners of the petition were away at official posts in other provinces and had no idea of the facts of the case. In the words of the magistrate, Huang "hid malicious intent under a fair countenance." Two strong magistrates had thus stepped in during roughly the last two decades of the Qing to provide leadership and, in the latter case, to counter Huang's expanding power. It seems clear from the Xiang Lake experience that if government officials were not susceptible to being suborned, they could rein in the arrogant power of local leaders. The nature and approach of the father-mother official (the magistrate) in the county seat was thus crucial in the nature and functioning of local society.

Huang Yuanshou would have another role to play in the struggle for the lake in subsequent years. It is striking that Yewu Mountain, which we have taken as symbol of the lake's imminent destruction, was central to or served as an important symbol in three of the four extant poems he wrote about the lake. Such poetic attraction seems to evince the role he tried to play in the lake's drama. In his poem "Reflection of the Clouds in Xiang Lake" Huang joins the scenic symbol of the lake with Yewu's image.[65] The decay of the lake is symbolized physically by the bay filled with marsh grass and poetically by the setting sun.

> The egrets fly up from the marsh-grass-filled bay.
> For ten thousand hectares the red setting sun shines in the porcelain lake.
> The reflections of the clouds shine in the mirrorlike waves.
> In the haze, like hair coiled in a knot, looms only Yewu Mountain.

The impending doom of the lake might have been averted if the Chinese government had possessed sufficient funds to dredge the lake properly and rebuild the dikes and sluicegates in a large-scale construction effort. But the Qing dynasty was reeling economically, not only from the

Taiping and three other midcentury rebellions but also from the onslaught of Western imperialist demands and wars (the Opium War of 1839–1842 and the Arrow War of 1856–1860) and subsequent indemnities forced on the Chinese government. There was simply no money for widespread reconstruction. The lake's imminent demise might also have been averted *if* a man the caliber of Gu Zhong or Wei Ji or Yu Shida had appeared to force issues and decisions on local government figures and to rein in the greed and corruption of yamen underlings charged with water-control duties or funds. But no such leader appeared, and the venality and arrogance of the sub-bureaucrats seemed only to increase. Finally, the destruction of the lake might have been avoided if the social structure had possessed some mechanism to articulate the public interest. When state officials abdicated, the problem fell to local leaders. In times of trouble and distress, the lineage- and family-centered nature of society mitigated against widespread cooperation in the public interest. When lineage leaders hunkered down to protect their own interests, nothing could give expression to the public interest or insure that it be carried out.

It is a difficult conclusion to avoid: like the Qing dynasty and perhaps traditional Chinese civilization itself, the lake's very existence was challenged by a convergence of political, economic, military, and natural forces in the late nineteenth century. Had the resources of the eighteenth century been available for the lake (and dynasty and civilization), scavengers of the lake (and foes of dynasty and civilization) may have been held at bay. It was not to be.

VII

VIEW

From One View Pavilion on Stone Grotto Mountain, 1925

CHAPTER

Fending Off the Scavengers, 1903–1921

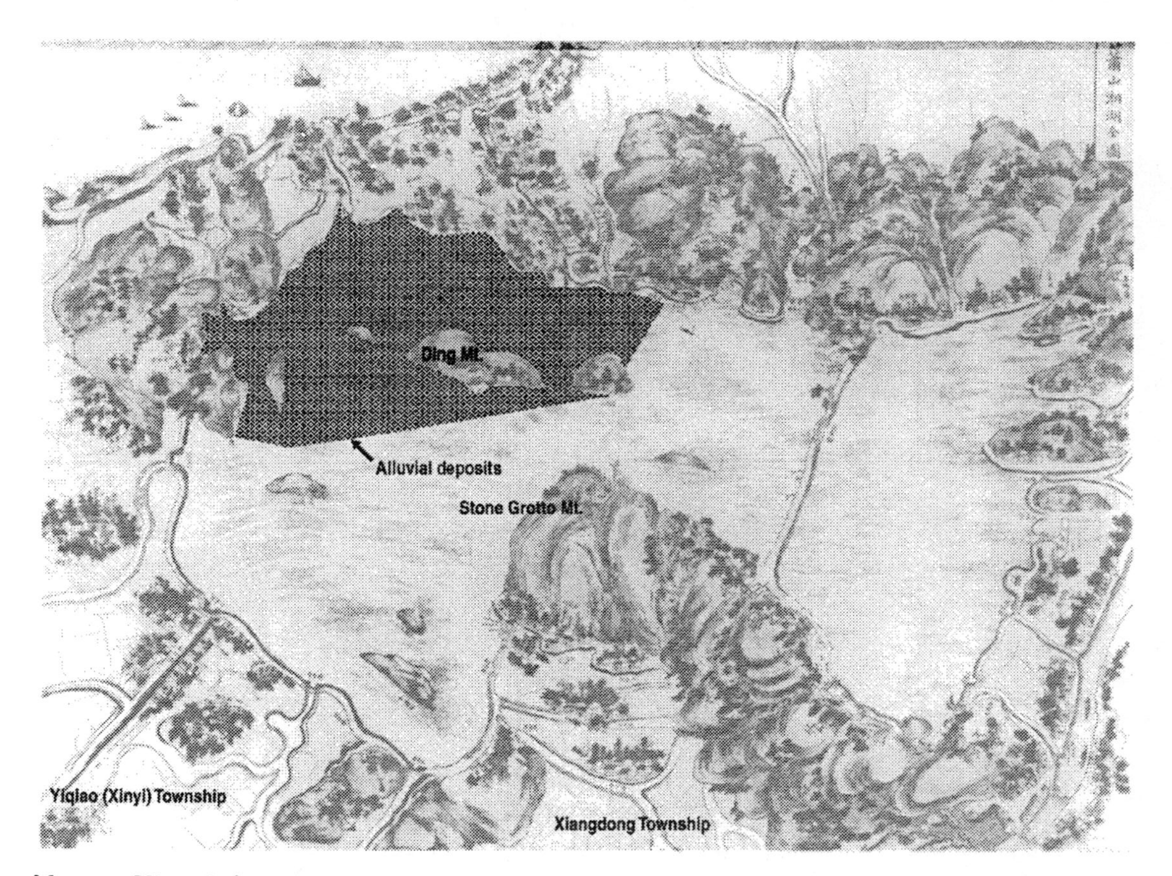

Map 11. Xiang Lake Sites, 1925

VIEW: *From One View Pavilion on Stone Grotto Mountain, 1925*

More renowned than the views from North Trunk or King of Yue Garrison mountains was the prospect from Stone Grotto Mountain, a 236-meter peak on the east side of Xiang Lake (figure 6). Throughout the centuries scholar after scholar from outstanding lake lineages had climbed the mountain to gaze at the play of sun and moonlight in lake's ripples; to marvel at the nearby mighty river, threatening (sounding "like thunder") and mysterious; to be moved by the unity of water and mountains in a seamless natural tapestry. The vistas and the emotions they engendered the men set down in verse.[1]

Much poetry about the mountain and the views from it evinces the glint of creative individuality based on each poet's personality or situation. Huang Jiugao, mid-sixteenth-century water and dike expert, wrote of water and dikes: of the water of Fragrant Spring at the top of Stone Grotto Mountain, of waterfalls cascading over the cliffs, of the pure, deep water of Xiang Lake, bounded by the natural dikes of mountains.[2] Buddhist Lai Sanping, late sixteenth-century jinshi, painted a picture of renewal and freshness:

> After the snow, the sky clears; the mountain looks new.
> All around are a thousand peaks like rugged jade.
> The cold has only deepened the green of the pine needles.
> The fragrance of the plum flowers fills the air.
> I follow the trail of footprints in the snow and finally come upon a Buddhist monk.

Figure 6. Reclaimed lake land, Stone Grotto Mountain in background (1986)

> Standing next to a brazier, he warms a jar of the first wine of spring.
> We sit in the light of the setting sun in silent contemplation
> While in the valley the steam from the tea keeps the bamboo company.[3]

Wang Congyan, late eighteenth-century scholar, depicted the mountain in feminine images, as a goddess with hair coiled high.[4] Tang Jinzhao, the official who in the 1830s would draft the government's severe antiopium regulations, returned home to mourn his father; on a day in late 1825 he climbed the mountain and wrote a poem in which he noted the continuing life and greenness of the pine and bamboo despite winter's death chill.[5]

It is likely that the men viewed the lake, river, and mountains from One View Pavilion, mentioned in much of the poetry and even used as a name for some of the poems. Built in 1531 by Shaoxing prefect Hong Zhu, it fell into ruin many times but was always rebuilt. Amid the area's general decay in the late nineteenth century, however, the collapsed pavilion lay unreconstructed among mountain weeds and wildflowers. Next to the ruined pavilion was Fragrant Spring, about four feet square and one foot deep; Hong Zhu had carved its two-character name in stone at its side. A Buddhist temple constructed soon after the creation of the lake itself sat nearby. Like the pavilion, the temple had seen various incarnations: Brightly Shining Shrine, First Light Shrine, Stone Grotto Temple, and, by the 1920s, First Light Temple.[6] When it was repaired in the Guangxu period (1875–1908), there had been talk of also rebuilding the pavilion. But it was not until 1917, five years after the overthrow of the Manchu dynasty and the end of China's imperial system, that the pavilion was reconstructed. Its resurrection was ephemeral; in 1921 a tremendous windstorm destroyed the pavilion.

In spring 1925 the wife of Lai Yinlian (née Jia) of Long River came to Stone Grotto Mountain to worship at First Light Temple. Now in her sixties, as a youth she had pledged to serve the Buddha through her actions. She saw the collapsed pavilion and as part of her service funded its reconstruction. By early July the work was finished, again providing a special place from which to view the vistas that had allured past men and women. Huang Benlun, who left a record of this reconstruction and who would the next year be active in efforts to prevent the destruction of the lake, noted the views now possible: the beautiful mountains in Shaoxing to the east, the flowing river to the west, and mirrorlike Xiang Lake stretched out below, the mountains in Upper Xiang Lake floating islands in the blue expanse.

Huang Yuanshou, son of anti-Taiping leader Huang Zhongyao and principal in the Hengzhu Dike case of 1901, mentioned Stone Grotto Mountain in several poems. His treatment of it, like his poem on Yewu Mountain, seemed to imply threats to the lake. In "Miscellaneous Songs on Xiang Lake," Huang mentioned the mountain primarily in relation to the lake sluicegate at Stone Grotto's southeastern foot. Noting that this was the sluicegate at the lowest altitude of all, he speculated that if it were left open, all the lake would drain out, leaving only dry land.[7] Even more ominously suggestive is the ending of a poem entitled "One View Pavilion" of unknown date; in it, Huang refers to the mountain by its common local name, Gander Nose:[8]

> Grasses grow on the lofty Gander Nose.
> This single pavilion encompasses ten thousand images:
> The flowing river, separating Wu and Yue;
> The stars in space;
> The mists rising at the foot of the mountain;
> The moon's reflection dancing in the heart of the waves:
> It seems as if the lake is filled with beautiful pearls fixed by bamboo on lotus leaves,
> As if the lake were filled with sparkling grains of rice tossed into it.

This poem evokes disturbing thoughts for an upholder of the lake: Huang formed the penultimate line using the first characters of five of the main islands in Upper Xiang Lake. Seen thus, this display of poetic virtuousity would read:

> Mei, Zhu, He, Ruota, and Ding mountains
> Are like grains of rice thrown into the lake.

It is almost certainly intentional that several of these mountains lay in the area silted up by the mid-nineteenth-century floods and that Huang chose to denote them by the image of consumable rice. The rice image and the image of pearls also inform each other. Both rice and pearls can be viewed as sources of wealth, commodities to be bought and sold. When these images are linked to the island names, Huang describes not primarily scenic vistas to be admired but parcels of land as sources of wealth to be reclaimed and sold.[9] It was a view that became increasingly alluring in the first decades of the twentieth century.

CHAPTER: *Fending Off the Scavengers, 1903–1921*

By the first years of the twentieth century, a sense of humiliation at China's weakness in the face of aggressive imperialism, especially military losses to the French (1885), to the Japanese (1895), and to the Allied intervention in the Boxer War (1900), had convinced many urban elites that sweeping change was necessary for China's survival. Traditional forms, ideas, and ideology, and especially the non-Han Chinese Manchu regime, became fair game for examination and criticism. The short-lived reform effort of Kang Youwei—to change China by an institutional revolution from within the government in Beijing—was snuffed out by the Empress Dowager's coup in September 1898; but the sense that something in the universe was askew and that basic change was essential could not be so easily extinguished. The Empress Dowager's own disastrous flirtation in 1900 with the Boxers as antiforeign saviors risen from the masses only deepened the sense of national abasement and renewed the drive to revivify China. In the first decade of the century, fundamental educational, military, and political reform became the agenda not only of concerned nonofficial elites but of the government, even the Empress Dowager herself. The civil service examination, which had produced not only the officials who served the emperor but also legitimacy for the social elite at all levels of the polity, was abolished in 1905. A modern school system was established in its place, but it lacked the ideological sanction of the examination. A modern army was built to bring China the strength

Figure 7. Silted Upper Xiang Lake near Yewu Mountain (1920s)

it needed to fend off future foreign aggression. In 1908 the government adopted a plan to establish a constitutional monarchy with representative bodies at township, county, provincial, and national levels. Both Chinese and foreign observers attested to the new hopeful spirit sweeping the urban and coastal regions of the country.[1]

The Reclamation Proposals of Huang Yuanshou, 1903 and 1910

Two years after the Hengzhu Dike case, Huang Yuanshou set forth a bold proposal to reclaim for paddy a large portion of the silted land near Ding Mountain (figure 7).[2] It was the first formal proposal to reclaim the lake since 1119. In intervening centuries, powerful figures—whether officials like the Song dynasty's Intendant Zhang or Zhou Ren or nonofficial elites like the leaders of the Sun and Wu lineages—had simply seized areas of lakeland, controlling them on their own de facto authority. Huang's proposal provided a rationale to fit the spirit of renovation and reform, evidence of the changed context of the early twentieth century.

Couched in the values and language of the late Qing reformist effort, the proposal purported to be a method of strengthening the nation and enriching the people. These twin goals of wealth and power, widespread among political leaders in the growing nationalism of what has been called the "inner core" of economic development, could be attained in part, Huang posited, by emphasizing traditional Confucian desiderata: expanding the county educational system and increasing agricultural productivity. To accomplish these ends, he proposed the establishment of a modern-style corporation to be named, translated literally, the Honest and Upright Company (*houzheng gongsi*)—an ironic name, considering that the proposal evidenced inordinate greed on Huang's part. The company would reclaim over six thousand mu of lakeland: of the rent from the reclaimed land, 30 percent would be earmarked for educational purposes, the remaining 70 percent for company coffers. Huang sought to turn a substantial personal profit.

Since his father had been dike manager for many years, Huang must have been aware of the dozens of prohibitions over the centuries against reclamation and encroachment. He tried to assuage the opponents of reclamation by offering to dredge the remainder of the lake for better irrigation and to build a stone dike as a barrier to further encroachment after the project was completed. He also alluded frequently to an underlying rationale: reclamation would increase arable land for the densely populated area and would lessen the ancient contradiction between land and population. Huang petitioned Zhejiang's governor, who was clearly solicitous for the local inhabitants and their views.[3] The provincial treasurer deputized a special investigator to visit the site of the proposed reclamation and meet with local leaders. On-site inspection revealed that Huang had overestimated the reclaimable portion of the lake by about 20 percent.[4] Huang's background, which gave the lie to the pretensions of morality in his company name, was also at issue. Huang had been dismissed by an earlier governor for extortion in collecting the county cigarette tax, a misdeed that came to light in the petitions of farmers from the townships around Xiang Lake.[5]

Attempts at consultation with local elites showed some of the problems encountered by such figures as He Shunbin, Mao Qiling, and Lai Qijun: few elites would speak out against one of the area's most powerful figures. Intimidation in the face of authority also explains those hesitant and hedged elite responses that were elicited. Fortunately for irrigators, the investigator and provincial treasurer came down strongly against reclamation, joining officials from the preceding decade and a half who had put the brake on deteriorating lake conditions and encroachment schemes. The treasurer warned local leaders about men who in these changing times used "new methods" or "new institutions" (*xinfa*) as a convenient pretext to enhance personal power: in the name of public values they sought private gain. The governor, following his advisers, rejected Huang's proposal and, following tradition, ordered the erection of a stele prohibiting future reclamation.[6] The proposal thus had been crushed by strong official opposition; though officials had sought consultations with interested local elites, the reactions they encountered suggest hesitancy about speaking out against Huang.

In his decision the governor cautioned the leaders of the "nine townships" to rectify the lake's problems by dredging alluvial deposits, thereby staving off future reclaimers. This admonition was significant: after the first century of the lake's existence the sources rarely mention the nine townships' responsibility to act as a *community*. The irrigation community, by definition one system drawing water from one lake, nevertheless functioned like a number of separate systems attached to particular lake sluicegates and channels. Concern focused on one problem—a leaking sluicegate, a crumbling dike, the altitude of Xinyi Township, the Cross Lake Bridge—as if it alone were to be dealt with. This is the first irony: despite the Chinese holistic view of the universe, which stressed the interconnectedness of all things, the functioning of a mundane irrigation system was compartmentalized by the centripetal forces of locality, family, lineage, and personal connection. The closed nature of traditional Chinese society seemed to preclude the long-term health of the larger system, though clearly "muddling-by" for many years was possible. Only if mobilized or goaded to action amid a dire threat by a strong figure—such as Lai Qijun in the 1780s—did the townships temporarily become a community.

It is striking that the nine-township community became increasingly important in the twentieth century at a time when the Chinese, in their push to national cohesion, began to experience a birth of associational activity.[7] Like the late eighteenth-century emphasis on the public sphere, the emphasis on wider collectivities bespoke a new China. This, however, is the second irony: the reemphasis on the nine-township community came in the twentieth century when the practical irrigation community was composed (and had been since the sixteenth century) of only three townships and parts of two others. Only in the annual payment of additional taxes (agreed to in the Song to cover the loss of revenue that would otherwise have been felt by the lake's creation) did the nine townships share. The governor's appeal for all nine townships to dredge the lake would likely, then, fall on a number of deaf ears; four townships and parts of two others simply had no self-interest in doing so. The wonder is that since the sixteenth century none of the latter had broken the inertia and attempted to change the taxation system. The appeal, then, was anachronistic, an appeal to the pre-Ming lake community. It was a call to act according to the forms and institutions of the past, which no longer held relevance for the present. As such, it could never effectively arouse the community.

Although similar to his earlier scheme in substance, Huang's 1910 proposal was more acute politically.[8] He argued that the matter was not simply a question of preserving old regulations for their own sake. The times permitted the overturning of the centuries-old prohibition against reclamation because the world was accommodating itself to the spirit of change. Claiming that the criterion for accepting change was whether it brought advantage, Huang stipulated that the Qiyan excavation of the 1450s and the Three River Sluicegate construction of the sixteenth century changed the area's conservancy to the area's advantage. His reclamation plans, he averred, would do the same. To the new governor, Huang suggested that the 1903 rejection had come primarily because of native-place "connections," that the former governor had been persuaded by a Hangzhou official to protect the lake as a favor to one of his protégés from Xiaoshan. Whether this particular allegation of the role of native-place connections is valid, native-place ties remained a potent factor in many developments regarding the lake.[9]

To strengthen his proposal, Huang named eighteen leaders from the nine townships as copartners and managers of the proposed company; this shrewd move illustrated to Hangzhou that Huang had the support of important local figures. The list is interesting for what it reveals about Huang and his supporters. Judged by traditional scholarly credentials, it was an impressive group, including ten juren and two gongsheng degree holders. Three of the group had offered support for Huang in his 1901 dike caper, suggesting a core of long-time "connected" individuals.[10] The depth of their commitment to reclamation is, however, suspect: of the eighteen, eight (for whatever reason) opposed later reclamation schemes and only one supported a subsequent plan. One wonders whether they supported Huang's efforts or whether Huang was simply able to use previous connections to gather a list. Finally, Huang made his appeal in the name of the "nine townships"; but it is notable that fewer than half (eight) of his eighteen supporters came from townships still dependent on irrigation water from the lake.

When the provincial investigative deputy and the magistrate held a hearing at the county yamen in late November, they were, not surprisingly, met by silence; only about ten men (*shenmin*) attended. The silence's probable causes were Huang power (Huang Zhongyao was still an active octogenarian) and a false sense of security from the prohibition about reclamation just six years earlier. To their credit, these officials decided that this serious matter required discussion by a larger number of local representatives.[11] Meanwhile the provincial governor and treasurer received memorials of support for Huang from five powerful men in Hangzhou and Xiaoshan.[12] Whether it was simply the clout of these powerful figures or other unknown factors, the Hangzhou government, on the condition that Huang amend some company by-laws, gave Huang permission to proceed with his plan.[13] (It later came to light that support from at least three of the five was trumped up, a tactic Huang had used in the Hengzhu Dike case almost a decade earlier.)[14]

The decision to allow Huang to proceed brought an outburst of opposition from the lake region in April 1911. Led by gongsheng-degreed dike manager Han Shaoxiang of the powerful Yiqiao lineage, some four to five hundred men met at the county seat. Like Lai Qijun in the late eighteenth century, Han must have had the leadership qualities to incite such an interest, and his authority as dike manager was probably crucial. Those who attended were from only those townships that still used lakewater.[15] Discussion was heated and the meeting chaotic. Although he was able to maintain order, the magistrate could not persuade the men of the advantages of reclamation.[16] Opponents of Huang's proposal primarily attacked its substance, though several also made ad hominem arguments. References to Huang's previous corruption sparked insinuations that his reclamation project was simply a method of revenge exacted against his earlier impeachers. In addition, Han charged that Huang and his father, the former dike manager, had actually conspired to do nothing to correct the silted condition of Upper Xiang Lake and had adopted a policy of willful and malicious neglect in order to make a reclamation attempt more feasible. Though there is no way to determine the validity of the charge, as dike manager Han would have had access to the information on which such an accusation was based. The long years (1860s–1890s) of inaction in dealing with the lake's serious problems lend credence to the charge.

As for the proposal itself, opponents emphasized specific, practical objections, arguing that the traditional prohibitions on reclamation, far from serving as a straitjacket, existed to protect agriculture. Any reclamation, they claimed, would threaten the livelihood of the townships. Han pointed to 1785 and 1852, years of great drought, when there had been insufficient lakewater, and this before the disastrous flood of 1865 had deposited its silt. Furthermore, as argued in 1903, the reclaimed paddies would also require water. As for dredging, Han asserted: "We can dredge it a foot deeper, but we can't raise it an inch. How can we get water to drain out?" Because fields were widespread, he argued, machinery for that purpose was infeasible. In Han's words, the proposal was "filled with ten thousand things, but there was not one as it should be." Reclamation would be a tragedy for 25 percent of the county. Finally, in pointed reference to Huang's local dominance, Han alluded to the tragedy of Censor He Shunbin. The magistrate, though not accused of wrongdoing, was patronizingly blamed for following Huang without any knowledge of the situation. Huang was accused of scavengerlike greed, wanting to eat up the lake like a piece of meat.[17]

The trump for the local elites and farmers was the threat of social unrest. In a letter to Hangzhou authorities, Han pointed to the opposition that existed and to a government communication to Huang that suggested that the plan would be hard to carry out in the face of opposition. He asked how the government could dare ignore the private wishes of the local inhabitants.[18] (At least 642 men, both elites and commoners, from seven townships signed petitions opposing the plan.)[19] In his report of the meeting to the governor, the magistrate stressed the clamor and tension, ending with the warning that carrying out the proposal would probably ignite a serious social disturbance. He recommended delaying the project; the governor, noting the danger of disturbance, so ordered.[20]

The threat of social unrest had been used effectively in the Lojiawu quarrying case of the 1750s. Then the imputed threat had been bandits. In 1911 the threat was allegedly local elites and law-abiding farmers, mobilized once again by a strong leader, Han Shaoxiang. The difference was the time. By spring 1911 political unrest was growing; in some of China's major cities demonstrations beginning the preceding fall had called for rapid adoption of a constitution and full-fledged legislative bodies. That spring also brought rice riots to the Hangzhou area.[21] In such a volatile situation, officials were unwilling to destabilize a local area. It was Huang's last hurrah; illness prevented him from undertaking other proposals, and his father, perhaps his strongest asset, died in 1912.[22]

The lake region's late nineteenth-century silence ended with the perception that officials supported the preservation of the lake. The 1903–1904 government decision and attitude were crucial in providing a sense of protection from the potential threats of Huang and his group of supporters. Strong leadership by Han Shaoxiang gave the opposition direction, voice, and authority. Local leaders clearly showed their awareness of their newfound voice by their calculated threat of social unrest. Huang too had learned from the 1903 proposal that the government was listening to local leaders. He added prestigious local backers to his 1910 proposal, but still he did not succeed. As the moribund dynasty and monarchy drew to a close, lakeside elites had been revivified.

Local Efforts at Reclamation, 1912–1914

Xiaoshan county and Xiang Lake saw little fighting in the revolution of 1911–1912, which overthrew the Qing dynasty and established the Republic of China. To be sure, bandits took advantage of the unstable situation to steal and kill.[23] But after the establishment of a new county government, in which only a handful of leading local figures played a role, on the surface the situation seemed little changed.[24] For many area people, however, the abdication of the Manchu rulers and the establishment of representative constitutionalism meant the rise of a new, almost palpable spirit and the ratification of the forces of changes underway since the end of the nineteenth century.

The first years of the republic saw a continuation of active local elite involvement around the lake, not only as defenders of the lake but also as potential reclaimers themselves. In April 1912 ten men headed by leaders of the new local "self-government" bodies, a township manager and a township council chairman from west of the lake, proposed a reclamation plan.[25] The alleged purpose was to provide capital for relief for the poor. Arguing that the lake currently did not permit effective irrigation, the group contended that unless reclamation and dredging were carried out, irrigation would eventually become impossible. Referring to the recent rejections of Huang's proposals, the petitioners argued that since the revolution the national essence (*guoti*) had changed and that the spirit of the republic was spreading. The project would be directed by officials and managed by the gentry (*guandu shenban*) and would reclaim over ten thousand mu of the lake using farmers as laborers. Recognizing substantial opposition to reclamation, the plan called for soldiers to protect the laborers. It also advocated dredging and using collecting pools rather than lakewater for the new paddies. Seventy percent of the profit was to be divided among the nine townships for public expenses (the same percentage was to be company profits under Huang's plan).

The record shows substantial discussion and debate over this proposal in the lake region. The magistrate met several times with the nine townships' representative "self-government" leaders and asked for meetings among elites in their townships.[26] Investigation and conferences showed that local opinions ranged from strong opposition in Yiqiao (formerly Xinyi) Township southeast of the lake, to support from the west of the lake, to support with reservations scattered among the townships.[27] The reservationists did not want a company to control the work, fearing changing a public interest to private gain, and they called for the land to be divided into nine sectors, parceled out to the townships, with the profit used for local self-government funds. Meetings continued for over two years.[28] In August 1914 the provincial civil governor ordered the magistrate to convene a meeting of representatives of the nine townships to vote on reclamation. When the majority opposed it, the governor ordered the prohibition on reclamation to continue.[29]

Notable in this episode of local scavenging is the continued official assumption that the locality and its leaders should decide the future of the lake. It is also evident that, if no immediate outside developers or threats to the lake existed, the "community" of irrigators, now alert to challenges to the lake, split into smaller communities of self-interest. There was considerable sentiment around the lake for reclamation but disagreement over method and hesitation over its import. Finally, the ten reclamation proposers (with one exception) came from area lineages that were not recently among the area's outstanding, as measured in record of local service or civil service degrees.[30] In part the proposal may be evidence of relative "have-nots" attempting to better themselves at the expense of the larger community.

Local Self-Government Leaders and Direct-Action Scavenging

The delay in decision making about the local proposal did not prevent some individuals around the lake from taking matters into their own hands. From May 1912 to July 1913, several men in township self-government posts began to reclaim pieces of lake land for their townships—rivalry among township leaders was intense, and many questioned the public nature of Xiang Lake.

In May 1912 the head of the Xiangdong (formerly Chonghua) Township council asked the magistrate to devise ways for self-government bodies to direct land reclamation. He noted that self-government regulations gave township councils rights in determining their funding methods and that before the 1911 revolution the council had decided to reclaim plots of wasteland at the foot of Stone Grotto and Climbing Melon mountains to use the proceeds for its needs.[31] The magistrate responded two days later that the plots were lake land and that reclamation was not permitted.[32]

Men from Xiangdong Township apparently planted the land anyway. In late September the township manager reported to the magistrate that about three hundred armed men from Yiqiao Township had come to the reclaimed land and cut down the rice. Calling it Xiang Lake "public" rice, the Yiqiao reapers also seized boats loaded with harvested rice and headed for Ding Mountain Village. The manager of Xiangdong Township blamed the destruction on two leaders of the Yiqiao Township council, Kong Duanfu and Han Shuirong. He asked for a swift investigation of Yiqiao Township leadership, claiming that, although the spirit of the republican era was nonviolent, "self-government" had become "chaotic government."[33] The head of Yiqiao Township's council (surnamed Han) defended Kong, who at the time of the alleged offense had been buying elementary school books in Hangzhou, and Han, who had been involved in census-taking for upcoming elections. But he did not relent in the township's position on the Xiangdong actions. He argued that any produce from the land was public because the reclaimed land belonged to all townships. He condemned Xiangdong actions as especially audacious in light of a current drought. In this small battle over lake land, the magistrate blamed Xiangdong for its reclamation but contended that Yiqiao had not used the proper method of redress.[34]

In May 1914, in the process of constructing a sluicegate on one of the lake's dikes, lakewater was accidentally released, with the result that some fields lacked sufficient water for crops. Some farmers went to the edge of the lake and began to farm land there. They armed themselves and had to be removed forcibly by police. In his initial report the lake dike manager blamed the action on local villains (*tupi*), a clear indication that some decision makers had little understanding of the plight of area farmers. The situation had polarized. Anyone who proposed reclamation or took it on themselves were, in the eyes of local upholders of the lake, "local villains"[35]—a marked contrast to the greater understanding and humaneness of late eighteenth-century approaches in dealing with petty encroachments. The scale of these earlier incidents also sets them apart from Huang Yuanshou's efforts and the area elites' attempt of 1912. The crucial distinction in all the reclamation activities and plans, however, was between the public and private realms in matters concerning the lake.

Nonlocal Developers: The 1915 Proposal

A proposal by two nonresident developers in autumn 1915 brought this concern to the fore as a full-blown issue.[36] Although the matter of the public versus private spheres had been clearly set forth first in the eighteenth century, the question as it related to the lake's legal status became significant only in the early years of the twentieth century. Into the rhetoric came a concern for legal rights, brought by Western approaches and significant because of expanding coastal capitalism and a rapidly burgeoning emphasis on law. Through the centuries changing categories framed the nature of the struggle over the lake: from the Song to the late seventeenth century, traditional morality; from the eighteenth century to the fall of the monarchy, specific pragmatic concerns; from the early years of the republic, legal rights and principles. Whereas lake preservers and reclaimers may have appealed to categories of morality, pragmatism, and legal rights, they continued to operate in a society based primarily on personal connection.

The proposal of 1915 appealed to recent government guidelines that allowed abandoned and newly formed alluvial land designated as national wasteland (*guoyou huangdi*) to be reclaimed. Specifically, the developers argued that the silted lake land was "official" (*guan*), not privately owned (*min*), and that it was "national" (*guoyou*), not "local" public land (*difang gongyou*).[37] They asserted that national regulations governing reclamation and development applied here, and they petitioned the Ministry of Agriculture and Commerce's concurrence.

The government responded by sponsoring a survey of the lake and alluvial area. The resulting report noted the obvious advantages of reclamation in an area that produced insufficient rice for its population. It also pointed to disadvantages: it would spark disputes over irrigation water and, more serious, would make it more difficult to protect the lake from further encroachment.[38] The reclaimable land amounted only to about three thousand mu, not the seven thousand suggested by the developers.[39]

As directed by the governor, the magistrate convened a meeting of county and township leaders (*shishen*) in December to discuss the proposal. The forty-one men who attended seem generally representative of the late Qing, early republican reformist elite-managers: all had served as self-government leaders since 1912; one was a provincial assemblyman; five held upper degrees from the traditional examination; and at least seven had been active in educational reform in the last decade of the Qing. The group unanimously opposed the plan, arguing that reclamation would simply mean insufficient irrigation for the townships. It also entered the arena of debate opened by the developers: the lake land belonged to the people of the nine townships (*mindi*), which had assumed the tax burden of the former cropland transformed to lake in the Song. Since these taxes had been paid for over eight hundred years, it was inaccurate to interpret the lake land as "official" wasteland. In addition, the group condemned the outside status of the developers, only one of whom came from Zhejiang.[40] They concluded that the plan was an effort of the greedy to swallow the lake and argued that government approval of the proposal would eliminate the rights of local leaders to decide local matters.

But the structures for decision making had begun to change. A specialized institution established in 1913—a provincial conservancy council—now had authority in this area. The council had decided on a joint program of dredging and reclamation, but the amount to be gained from the sale of new land was almost 200,000 yuan less than the estimated cost of dredging. The new provincial governor, much less supportive of reclamation than his predecessor, thus turned the decision over to the provincial assembly in late 1916.[41] The assembly accepted local arguments, deciding that the lake land could not be "official" land because the townships had paid almost eight centuries of taxes. It could not, therefore, be bought by developers without the approval of township leaders.[42] Reclamation should be prohibited in order to preserve irrigation. Once again the lake was protected by an official, this time in collaboration with the four-year-old representative provincial assembly. Despite the assembly's tax-based decision, it is clear that had it not been for budgetary considerations, outside developers would have received permission from the conservancy council to proceed with reclamation. The lake's preservers could not have rested easily after the episode of 1915.

The Dynamics of Opposition to Reclamation

In March 1921 the brother of Huang Yuanshou, Huang Pengshou, petitioned to reclaim the alluvial area and dredge the lake for better irrigation.[43] Huang had apparently learned from his brother's abortive efforts to reclaim the land. He instituted public discussion of his plans *before* his formal petition, and his project was to be managed by the nine townships, with the monies to be used by the townships for public affairs. Huang's hopes were cut short almost immediately; that month representatives of the townships requested prohibition of any reclamation or dredging scheme.[44] Fifteen men, called "representative citizens of the nine townships," opposed Huang's project through their petition. Analysis of this opposition suggests the dynamics of resistance to reclamation in the early republic. Ten came from Yiqiao Township (formerly Xinyi and Xuxian townships) southeast of the lake. Of the five townships that still received lakewater, only Yiqiao depended completely on Xiang Lake. Townships to the west of the lake had other small lakes from which to irrigate, and Lao Lake Village in former Yuhua Township only occasionally required lakewater. Residents of Yiqiao would, therefore, be especially concerned if Upper Xiang Lake, the source of water for their crops, were substantially changed. Self-interest was crucial in the opposition.

Another was the preeminence of the Han lineage: five of the ten from southeast of the lake who opposed Huang were Han leaders. The Han tried to protect their livelihood—landholdings as well as commercial concerns. In prime position to help preserve those interests and the lake was one of their lineage leaders, lake dike manager Han Shaoxiang. Constantly vigilant to any threat to the integrity of the lake or lake facilities (such as dike damage caused by drying and hauling water weed fertilizer),[45] Han conscientiously used the annual repair fund for routine maintenance of dike and sluicegate, in sharp contrast to comparable leaders of the nineteenth century. When in 1919 floods brought widespread damage, Han carefully saw to the repair and reconstruction of affected facilities.[46]

The prestige of the lineage and of Han Shaoxiang himself was enhanced further by the important role lineage member Han Shaoqi played in the provincial military hierarchy. In the 1911 revolution Han Shaoqi had been one of a handful of leaders in Hangzhou; afterwards he remained close to the center of power.[47] (The Lai lineage at Long River near the lake also had one of their own, Lai Weiliang, in a similar military position, and prestige accrued to that long-time powerful lineage.) Because the power of the provincial government in the early republic was greater vis-à-vis the central government than it had been during the empire, and because during the empire the law of avoidance made it impossible to serve in official posts in one's native province, the situation of officials from Xiaoshan serving in significant provincial positions across the river in Hangzhou was unprecedented. As always in the Chinese context, lineage and personal prestige brought by such connections translated almost directly to greater stature and more power in day-to-day issues.

The identity of the reclaimers was also important in the rise of opposition to reclamation. Generally speaking, if preservers of the lake had close ties to the most powerful area leader, reclaimers had little chance of success unless they had connections with powerful officials who might be persuaded or bribed about the wisdom of a planned project. From 1903 to 1921, to greater and lesser degrees and for varying reasons, officials championed the cause of the preservers of the lake, and prestige and power in the Xiang Lake environs centered in lineages that were upholders of the lake. Even more, would-be reclaimers lacked the status needed to contend for lake control. They were representatives of what might be called second-rate lineages (in terms of local power) (1912–1914). They were outsiders, who could always be expected to face indigenous opposition in the particularistic closed social system (1915). And they were such men as Huang Pengshou (1921), whose family had lost considerable "face" (prestige) because of both his brother's perceived duplicity and greed and his father's and brother's supposed conspiracy to destroy the lake. At a time when entrepreneurs had begun to reclaim lakes for investment in many areas, it is not surprising that the opponents of the reclamation of Xiang Lake were successful in their aims.

A final aspect evident in dealing with lake encroachers in the early republic was the violent, excessive reaction of local leaders and commoners to encroachment by obvious social inferiors. The reaction (with three hundred armed men) of Yiqiao Township to the limited planting of rice by Xiangdong Township denizens in 1912 is a case in point, as is another affair of 1912 around Ding Mountain. There a man had taken advantage of the silted condition around the island to encroach on about one hundred mu of land. Over one thousand residents of Upper Xiang Lake angrily went to the area, attacked the encroacher (called a "local villain," or *tupi*), and pulled up and destroyed all his rice. In a letter describing both this affair and a minor encroachment attempt in 1917, the writer noted several times that "when the masses are angry, it is difficult to violate [the lake]"(*chongnu nanfan*).[48]

The story of the lake, however, as we have seen, is one of docile masses, kowtowing before prestigious and powerful lake destroyers; only if goaded or supported by a strong official or a local leader as prestigious and powerful as the reclaimer would the masses or even other local leaders act. They would act, with great anger, at those who had no claims to prestige or "connections." The lake culture often seemed marked by dual attitudes: fear of and fawning on social superiors, and contempt of and preying on social subordinates. For that reason, the heroes—those who tried to stop the predation—tower from the pages of the past as truly extraordinary individuals.[49]

VIII

VIEW

From One View Pavilion on Genuine Peace Mountain, 1937

CHAPTER

Ends and Means, 1926–1937

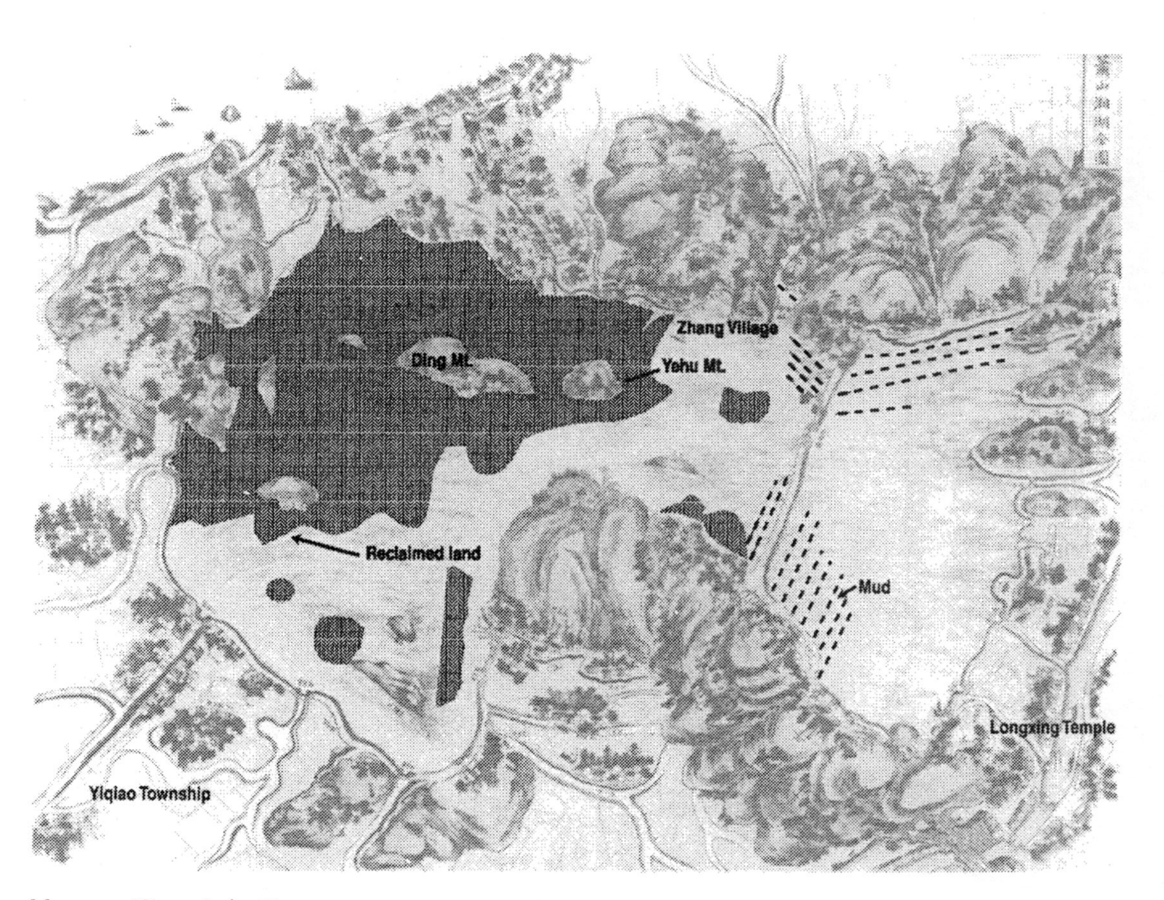

Map 12. Xiang Lake Sites, 1937

VIEW: *From One View Pavilion on Genuine Peace Mountain, 1937*

It is the same pavilion on the same mountain as twelve years earlier; only the name of the mountain has changed: Stone Grotto to Genuine Peace. We do not know who changed the name or why; presumably the renamer believed Genuine Peace had greater political or ethical or religious import and was thus a "better" name. The change seems innocuous, though more purposeful than the surely sloppy phonetic twentieth-century metamorphosis of Yewu (Thrown into the Crow [River]) Mountain to Yehu (Thrown into the Lake) Mountain. The latter change, however explained, was anachronistic: Fan Zeng allegedly threw the mountain peak down almost a millennium before the lake was created.

Names have served many purposes in the history of Xiang Lake. We have seen the use of He the Filial Son, Censor He, and Mother Hu the Chaste Widow to designate human beings by title, social role, or attribute in order to depersonalize and symbolize. Names served to remake the old and symbolize the new. China became a republic with a president rather than an empire with an emperor—but day-to-day reality varied little. In 1909 the names of all Xiaoshan townships, which had existed since the Song dynasty, were changed with the establishment of representative self-government. Here reality changed with nomenclature: the nine were broken into parts of sixteen.[1] Yet the name "nine-townships" continued to be used for all lake matters.

Names obfuscated, rationalized, propagandized, and distorted. Starving farmers who tried to eke out a marginal life on a few mu of land at the lakeshore were tagged "local villains"; local leaders who tried to prevent large-scale destruction of the lake by reclaimers became known in 1926 as "local bullies and evil gentry"; in obvious greed Huang Yuanshou established his Honest and Upright Company. Depending on one's point of view, "reclamation" stood for progress or destruction, "progressive" signified positive change or subversion. "Official" (*guan*), "public" (*gong*), and "private" (*si*) became tangled and confused from blurring the concepts of ownership and land usage. Meanings of words and names were always relative to the speaker and often depended on individual understanding. "Reconstruction" in the 1930s and "liberation" after 1949 propagandized a particular image of what was occurring, the image of the imagers—the reconstructors and liberators. But it was always possible, of course, that these epithets were Honest and Upright Companies on a nationwide scale.

The view from atop Genuine Peace Mountain in 1937 was as different from the view on Stone Grotto Mountain twelve years earlier as if they had indeed been two different mountains. At the foot of Yewu Mountain in the rebuilt Clouds of Xiang Temple and on Ding Mountain were the facilities of the Xiang Lake Normal School. A large section of Upper Xiang Lake was now dry land with paddies, trees, and buildings: homes and sheds stood where once the water had been over fifty feet deep. Crops grew in soil where once the water-shield plant had flourished in luxuriance. Because names are significant, even if their relationship to reality is tangential, it is important to note the change of the name of the mountain, former island, around which the reclamation had occurred. Old and new are both pronounced "Ding," but the meanings differ. Until the late 1920s Ding was written 定, a character referring to one or two stars in the northern constellation of Pegasus. At the time of and after reclamation, Ding came to be written 錠, the radical to the left meaning "gold" and the new character having the meaning of an ingot, perhaps of gold or silver. From the heavens to the lure of wealth: the metamorphosis of a name.

Ding Mountain became the focal point of turning the water into land. It is certainly symbolic that the centuries have left no poetry about Ding Mountain. If poetry was a hallmark of traditional Chinese culture, the lack of poetry about Ding Mountain distances the mountain from that traditional culture and from the images set down in the lake poetry, images of natural beauty and plenty. Unworthy of poetic notice for centuries, Ding Mountain now became a symbol of promised wealth and gleamed like an ingot waiting to be processed.

There were, of course, other vistas from One View Pavilion. To the south and southwest, paddies formed a chessboard—in the summer, water-logged plots bounded by low dike borders and walkways; in the fall, field on field of yellow-gold rice. Spring, with its profusion of bright yellow rape flowers, vied with the golden ripeness of autumn, as if to clothe the countryside in ancient imperial yellow. To the west the Qiantang River still flowed. A 1933 county gazetteer noted that from the pavilion the sails of the boats plying the channels looked like seagulls in the water.[2] In early 1937, a railroad bridge across the river was finally completed about three miles upstream from the old ferry route between Xixing and Hangzhou. For the first time, the Shanghai-Hangzhou-Ningbo Railway could run the complete journey denoted by its name. Within months of its completion, however, it was destroyed in the devastating Japanese invasion in late 1937.[3]

To the north-northwest, Lower Xiang Lake wore a strangely mottled look.[4] Each year the brick and tile industry dredged tons of clay from the lake's bottom; each kiln took an estimated 200,000 cubic feet per year, and in the early 1930s there were sixty-three kilns. Dredgers dug pits many meters deep. Around these pits the mud gradually hardened into square-to-round dikes, giving the appearance, from the height of One View Pavilion, of fish scales. When rains came and covered these dredging "dikes," the irregular lake bottom made for treacherous boating, an undertaking made even more difficult by the mud flats built up to near the surface on both east and west shores.

The lake had become ponds of varying size: in Upper Xiang Lake, they formed the leftovers from the reclaimer's dinner, which had been "a fat piece of meat," in the words of one lake investigator.[5] In Lower Xiang Lake, they were the legacy of the brick and tile industry. And yet the name "lake," like "nine townships," remained, an anachronism whose use could not change reality. Though the work of the lake's heroes over the centuries was largely being destroyed, the fiction of its existence was not allowed to die. It was similar to Chiang Kai-shek's resurrection, three years earlier and amid increasing authoritarianism, of the cult of Confucius: persisting in past traditions and upholding the names of the past though they bore little relevance to current reality.

A tile pit covered by waves of lakewater.
It is there that fishermen find the delicious fish,
Especially when the peach trees blossom and lakewater rises in the spring.
Then in villages everywhere one can hear the calls of fish sellers.

—Huang Yuanshou, "Miscellaneous Songs on Xiang Lake," early twentieth century

Sun Quwen of Xiaoshan county recently died after eating fish that he had caught near Cross Lake Bridge in Xiang Lake. His children were also poisoned and remain in serious condition.

—Shenbao, a Shanghai newspaper, December 4, 1929

CHAPTER: *Ends and Means, 1926–1937*

In 1926 the bulwark against predation, the officialdom, turned predator. That April the Xiaoshan magistrate, Guo Cengzhen, proposed to reclaim the silted area to increase rice-producing land and help the county's poor.[1] Notable as the first reclamation proposal initiated by a local official in the lake's long history, this effort openly pitted officials against many in the region and ignited a public protest that, though ultimately futile, reached a considerable level. When Guo announced his plans, he called for township leaders to discuss the project, though he had clearly determined to proceed with the project. An assistant surveyed the area and drew reclamation boundaries in April and May *before* any discussion meetings were held.

In little more than two weeks after Guo broached the plan, anti-reclamation petitions from local leaders, who had been energized by official support since the last years of the Qing dynasty, began to arrive at the offices of Governor Xia.[2] They argued that Xiang Lake was the private property of the people (*renmin*) of the nine townships who had borne the tax burden since the Song: therefore the townships, not the county as a whole, must consent to any reclamation efforts. The petitioners emphasized any reclamation would endanger future irrigation. They charged that powerful bullies (*hao*) motivated by personal greed had planned the project. Careful attention to protecting irrigation through dredging alone would allow agriculture to prosper (*li nongtian*) and would pacify the masses (*an xiaomin*).[3]

Responding to an order from Governor Xia, Guo called a meeting on May 20 for citizens (*gongmin*) of the townships. Seventy-three men whose identities and native places are unrecorded attended and cast ballots to decide the issue.[4] Sixty-three voted to allow reclamation and dredging; eight proposed dredging alone; only two supported a prohibition. The decision did recognize the lake as the property of the nine townships: each township was to organize an office and choose three deputies to prepare for the dredging and reclamation. It appeared, then, that the project, conceived by the magistrate—and, most certainly, interested outsiders—would proceed.[5]

Petitions flooded the yamen after the vote charging that those who had attended the meeting did not represent the townships' feelings, that they had been motivated by the prospect of personal gain.[6] These petitions made one new argument: the lake was the townships' inheritance from the past and therefore could not be changed. This appeal to tradition was a clutching at straws. Many opponents of the new plan had opposed earlier attempts, but this was the first time that they had battled officials. This grave new stage of the lake struggle explains this appeal, which could not possibly have been taken seriously given the changes that had occurred to the lake in the preceding centuries. In many ways, it recalls arguments raised by nineteenth-century Chinese intellectuals and officials who, threatened by change, sought refuge in the past to escape the problems of the present. In contrast, Magistrate Guo appeared the concerned pragmatist: Xiaoshan, he argued to Governor Xia, was densely populated; it needed more land to produce rice in order to reduce the continually high price of this basic commodity.[7]

In his response to the magistrate's report of the meeting, Governor Xia, a Zhejiangese, took issue with the conclusion that the lake was the townships' property. Only in the late Qing, he contended, did certain men begin to talk of the lake in this way. Claiming that gazetteers recorded no such language, he accused wily opponents of reclamation of trying to thwart legitimate provincial rights. The character of the land, he asserted, was not in the least questionable: it was "official" land, and it could be reclaimed without agreement from any local leader.[8] The dredging and reclaiming of the lake were to be controlled and managed completely by officials and the county bureaucracy.[9] Such outright disdain of the locality and its leaders is remarkable and must have devastated the leaders of the antireclamation movement.

In the first quarter of the century, traditional social, political, and ethical bulwarks had been swept away one after another by the tides of change. The end of the civil service examination system had destroyed the traditional qualifications for official position and social status and had opened them to whomever might seize them—military leaders, burgeoning capitalists, bandits, scholars. The end of the dynasty cast China into a context it had not experienced for two thousand years: from an empire with established institutions and methods to the vicissitudes of an untried republic. And the era known as May Fourth attempted to cast off the old familial ethical predators, specifically the "superiors" in the so-called Confucian bonds: husbands, fathers, and elder brothers and much of what was seen as undemocratic and unscientific in the old culture.[10] The Chinese were searching for new ways, political, legal, ethical, social, and cultural. In an era that lacked a sense of restraint, many unconscionable predators seized for themselves. Predation that traditionally may have been tempered by a restraint imposed by ethics and established forms and procedures could become boundless. The face of predation could vary, but at Xiang Lake it was without question the face of the yamen. The implicit compact of the informal cooperation of state (officials) and society (local elites) was severed. Those in the county and provincial offices now abandoned even the guise of paternalism and preyed on local society—elites, commoners, and resources alike.

The Xiang Lake community, having been awakened from its nineteenth-century silence, was not about to be muted at this time. Public protest grew. At a general meeting of the nine townships' citizens (*gongmin dahui*) on June 19 a telegram was sent to Sun Chuanfang, military overlord of the province, asking for the proposal's rejection.[11] Signed by more than 1,300 people, it was followed by a letter to Magistrate Guo signed by over 1,500 citizens.[12] The latter, with copious references to decisions of historical personages from the Ming and Qing periods and specific data on taxes, tried to prove that the lake was owned privately by the nine townships. It also attacked the governor as a man from Zhejiang who would not look after the needs of Zhejiangese, clearly evidencing their belief that Xia was colluding with outsiders. The magistrate turned a deaf ear, replying that the decision was irrevocable.[13] The governor responded with direct quotations from historical records to refute the points of the opposition's brief.[14]

During these months the efforts of the lake community were bolstered by the support of two county men who were powerful and respected in provincial circles. Neither had been previously involved in lake issues, but the gravity of the situation compelled them to remonstrate through petitions. Wang Xieyang (juren, 1893) offered a general overview of previous reclamation prohibitions, an antireclamation brief, and a proposal for dredging the lake to be directed by denizens. The few for their own profit, he asserted, would affect the lives of many if Guo's project were undertaken.[15] Another more specific challenge came from Zhang Tao, a Xiaoshan provincial assemblyman and lawyer with political ties to Sun Yat-sen's (and by mid-1926, increasingly Chiang Kai-shek's) Nationalist party. Zhang challenged Xia's interpretation of the term *official.* Before 1912, Zhang argued, all land was either private (*si*) or official (*guan*); but since then, the term *official* could mean many things: city-owned, township-owned, county-owned, province-owned, nation-owned. He asked Xia for evidence in the case of Xiang Lake that *official* meant province-owned. When confronted directly, Xia responded not by answering but by emphasizing practical considerations. If the lake were not dredged and a portion reclaimed, in a number of years the entire lake would silt up, making irrigation impossible and thus diminishing the area's wealth. In addition, reclaimed land would increase the county production of rice by several hundred thousand piculs, which would relieve agricultural distress.[16]

In mid-July lake leaders sent Governor Xia a three-point impeachment of the magistrate for harming the townships' livelihood.[17] They focused on the age-old prohibition against reclamation; the taxes still assessed on the townships (if the lake was indeed "official" land, they argued, the townships should be repaid for over four dynasties' worth of taxes); and the provincial assembly's 1916 decision, which, they averred, carried the force of law. This appeal to law as established by a representative assembly was an obvious incongruity when arbitrary military authority from Hangzhou and corrupt civil rule from Xiaoshan were facts of life. The appeal reveals not only the lack of a realistic comprehension of the situation but also the extent of departure from such traditional procedures as reliance on personal connection. Guo responded that he was simply carrying out the governor's orders.[18] Many telegrams followed calling for the abolition of the already-established reclamation office. Xia ordered that the author of a strongly worded telegram be detained for slander and that efforts be undertaken to locate the whereabouts of its 371 cosigners.

Demonstrations erupted in Xiaoshan and Hangzhou. In the capital at least 750 people paraded to the governor's yamen, asking that the reclamation order be overturned.[19] Records of the demonstrations come only from the proreclamation side: they admit that poor farmers and "country rustics" from around the lake participated but contend that local leaders had compelled and even bullied people to participate. Xia reported to Guo that the blame should fall on gentry-managers (*shendong*) who were seducing the illiterate.[20] Provincial police were ordered to arrest those instrumental in the affair.

The center of opposition, not surprisingly, was again the township and town of Yiqiao.[21] After further investigation, only two men were deemed worthy of arrest and eight more were warned about further incitement of disturbances or seduction of illiterates. All were named "evil gentry" (*lieshen*). To be arrested were Han Dichang, gongsheng degree holder who had supported Huang Yuanshou in the 1901 and 1910 cases, and Li Changshou, who himself had proposed the local elite reclamation of 1912. Former would-be reclaimers were, therefore, accused of bribing county "rustics" to oppose reclamation; because neither had played any roles after 1910 and 1912, respectively, it is hard to ascertain the source of their changed position. In any case, it is a reminder that changed contexts often found the same men on different sides of the same issue. Those to receive warnings were the manager of the native bank in Yiqiao, which allegedly provided money for the bribes; the former council chairmen for Xiangdong and Yiqiao townships who had found themselves on opposite sides of the rice encroachment case of 1912; Han Shaoxiang, the conscientious preserver of the lake; three other leaders of the Han lineage; and a powerful merchant named Li. It seems clear that the sobriquet "evil gentry" could be acquired like later epithets—"Communist" or "capitalist-roader"—simply by opposing the "line" of those in authority.

Judging the validity of the allegation of bribery to incite the demonstration is impossible. Given the general nature of Chinese relationships to authority, bribes may not in this case have been necessary, because the leaders of the movement had substantial local prestige and authority of their own. What is surely the case is that farmers and "illiterates" did not travel by ferry to Hangzhou without the encouragement and support of the leaders; such independent action would have flown in the face of every action of Xiang Lake's inhabitants through the centuries. The suggestion that they did not act alone, however, does not mean that they were unaware of personal advantage or disadvantage in the situation; it simply points to the hesitancy of independent action in the face of authority and without the support of counterauthority.

There is no information on the outcome of the arrests or warning orders, but protest letters and demonstrations ceased soon afterwards, suggesting that the arrests, as part of Xia's efforts to stifle unrest, were successful in achieving his objectives. It is also likely that a disastrous flood in August turned the focus of many from the future of the lake to mere survival. The West River Dike and the North Sea Dike broke in ten places after heavy rains: 44 people were killed, almost 1,400 homes were destroyed, and 132,700 became refugees.[22] Whatever the case, reclaimed land went on sale in Hangzhou in mid-October. A few days later Governor Xia was killed in a coup attempt to throw off warlord Sun Chuanfang's control of the province.[23] His successor, Chen Yi, a native of Shaoxing, ordered Magistrate Guo to forbid any further reclamation activity. Chen, who was probably moved by connections to his fellow native-place leaders, told Guo that Xia had overridden the opinions of the nine townships and had thereby overturned the accomplishments of past generations. In December the magistrate ordered dike manager Han Shaoxiang to inscribe a marker forever prohibiting reclamation.[24]

The events of late autumn 1926 clearly point to what would later come to be called "politics in command." Policy changed as quickly as political regime gave way. The power of connections, here native-place ties, changed the direction of local affairs in an arbitrary and, for defenders of the lake, fortunate manner. The method of the change of policy, however, could not have left the proponents of irrigation with much optimism. The episode was the beginning of a trend for later twentieth-century environmental policies: the sponsoring, or at least encouraging, of the reclamation of lake land by state officials, a trend in sharp contrast to what had been the state attitude toward Xiang Lake for centuries.

Between 1903 and 1926 the political status of local leaders and the legitimacy of local political power relating to the control of Xiang Lake changed perceptibly. The power of local leaders so recently reawakened was rapidly diminished by arbitrary official power. In this period of political flux, official and nonofficial elites sought some firm political legitimacy to anchor their arguments. Such words as *public* (*gong*), *official* (*guan*), and *private* (*si*), while used in the past, took on new life in establishing legal definitions and became new parameters for answers to social and political problems. Developers could reclaim the lake because its legal status gave them the "right" to do so no matter what the local community claimed or wanted. The older consensual method of decision-making in meetings between officials and local leaders was being replaced by decision making grounded (at least in theory) in a more detached legal framework.

Finally, the validity of each side's positions must be judged. Obviously both reclaimers and lake defenders put forth their most persuasive arguments. No observers were neutral; the surveyors were not disinterested, hired, as they were, by the official reclaimers and reporting back to them. The reclaimers never effectively dealt with the irrigation issue: how would those dependent on lake water for their crops cope without the lake? Oddly enough, the defenders of the lake and irrigation themselves skirted this central issue, emphasizing instead the historical record and the centuries-long tax status of the land. In the end, the validity of arguments became meaningless in the face of the decisive factor of political power. Circumstantial evidence suggests that the official reclaimers were involved in the project for self-aggrandizement. Xia Chao was known as an ambitious, power-hungry military politician. And in April 1927, in a tell-tale act revealing his public standards, Magistrate Guo absconded to Shanghai with over 100,000 yuan of government money.[25] In addition, a report of investigators in the summer of 1927 put the onus of venality on officials and wealthy figures who wanted to have "their fingers in the pie" (*ranzhi*).[26]

Reclamation as "Reconstruction"

The lake's reprieve was short. In February 1927 the eastern army of the Guomindang's Northern Expedition seized Hangzhou, and the province of Zhejiang came under Chiang Kai-shek's control. The Nationalist victory promised to destroy the warlord power that had bedeviled China with its arbitrary militarism for over a decade; it held out the fervent hope of finally containing aggressive threats from foreign imperialists, who had gone from making war themselves to putting their money on warlords as surrogates; it raised hopes dashed after the 1911 revolution for the fulfillment of national unity and strength. Its agenda was "reconstruction," building national strength after years of turmoil. By late spring the Nationalist provincial party bureau proposed to develop Xiang Lake as the agricultural experiment farm (*nongchang*) of the Third National Sun

Yat-sen University's Workers' and Farmers' Institute.[27] Both the provincial and the central government approved, and the university, which was not formally established until August 1, sent a team of inspectors and surveyors to the lake in July.[28] The team included Mou Shusheng, one of former Magistrate Guo's confidants. Close connections between the developers in 1926 and those in 1927, despite the new "revolutionary" government, are a likely possibility. The team's report is therefore not surprising:

> Xiaoshan's Xiang Lake has fertile soil, abundant natural products, sufficient water, and beautiful scenery. In all the province it is the most suitable location to establish the experiment farm. . . . Moreover, to reclaim and dredge the remainder will produce absolutely no deleterious effects on the irrigation for the nine townships; in addition, the livelihood and culture of the people living along the lake will be greatly improved.[29]

The report noted that whereas the original lake's circumference was over 82 li, the circumference in 1927 was only a little over 56 li, smaller by one-third. The lake originally watered 146,868 mu of land; in 1927, it irrigated only 58,747, or 60 percent less.[30] Because of such a reduction, the report argued, the lake contained more water than was actually necessary. About 8,000 mu of lake land should be reclaimed, starting with the over 3,000 mu around Ding Mountain that was already above water level. Ding Mountain was described as an excellent site for such an undertaking: it had fertile soil, was near a plentiful supply of agricultural and reclamation workers (more than a hundred on Ding Mountain itself and many more in other lake villages), and was conveniently located for transportation (figure 8).[31]

The team, operating like the new government under Chiang Kai-shek after 1928, set down detailed blueprints with specific schedules as guidelines in this attempt to "revolutionize the farmers' spirit and [eventually] spread the spirit of agricultural experimentation throughout the province."[32] The experimental farm's aim was to improve a wide range of agricultural, aquatic, and timber products already grown in the area.[33] During the first stage, until mid-April 1928, the farm would be established, with reclamation beginning after workers were hired and trained. Before October 1928, dredging of transportation channels would begin and the tile industry would be put in order with improved methods of firing. By 1930 an electric-generating plant would be built, agricultural equipment purchased, and a library, agricultural exhibition hall, houses,

Figure 8. Yewu Mountain, left, Ding Mountain in background (1986)

and roads constructed. By the end of 1931, construction of a hospital, hotel, pavilion, and agricultural products factory was to be completed. The estimated total cost would be one million yuan.[34]

The impossibly ambitious schedule set forth by the investigative team was quickly discarded. The dredging of a river into the reclaimed area, for example, scheduled to be completed in 1928, was not undertaken until spring 1933. Though the experimental farm was begun in late 1927, it was not completed until 1934, by which time an agricultural extension office was also in operation. At Yewu Mountain the Xiang Lake Normal School had also been established and was increasingly renowned or infamous, depending on one's viewpoint. Over 7,000 mu had been reclaimed by 1934, with most (over 5,000) owned by the Reconstruction Bureau of the provincial government; over 600 mu was owned by the experiment farm, then controlled by what had become Zhejiang University; 100 mu was owned by the normal school (to which we will return shortly); and 400 to 500 mu were roads, residences, ditches, and private gardens.[35] Some in the provincial Reconstruction Bureau were pleased with the progress; by 1936 they were making plans to reclaim up to 10,000 mu for productive farm land. But that goal was halted by Japan's invasion.[36] Within a decade of the beginning of officially controlled reclamation, a good portion of Upper Xiang Lake had disappeared, and most of the rest was being slated for extinction.

And what of the people around the lake? One searches in vain for any response, much less protest. Newspapers, while reporting many violent episodes in 1927–1928, including Guomindang party purges of left-wing members, bandit disturbances, strikes, and similar social outbursts, do not mention unrest or protests around the lake. The provincial government gazette, which details robberies on elite homes even to the point of cataloging items stolen, mentions no disturbance at the lake or any petition for redress.[37] Periodicals covering the Xiang Lake reconstruction make no mention of protest by irrigators or by Yiqiao town or township, although they do cover other disturbances at the lake. One is met simply by the silence. Just as the arrests and threatened action against protestors in the summer of 1926 snuffed out further disturbances, it seems that the authority of the new government, bolstered by party surveillance and with active local police forces, had once again intimidated local leaders and their supporters.[38] It was the worst of all possible situations: With government as predator, what recourse did the defenders of the lake have?

The Costs of Reconstruction

The human cost of reconstruction was high. With the reclamation came many outsiders in search of personal advantage; one writer ironically (given the revolutionary political claims of the new regime) likened the large number invading the Ding Mountain area to the imperialists who invaded China in the nineteenth century.[39] Men seized Xiang Lake's famous old landmarks, occupied them, stole and sold them.[40] Developers and their hangers-on descended on the area like a new version of Taiping vandals, brandishing scientific instead of religious shibboleths.

Most drastically affected (apart from irrigators) were the workers in the brick and tile industry. After 1927, when tenants came to farm reclaimed land, which included some of the brickmakers' former dredging areas, farmers and kiln households frequently clashed. As contentions increased, litigation swept into open fighting, home burning, and widespread personal injury.[41] The provincial government reacted by prohibiting all mud-dredging in Upper Xiang Lake. It furthermore decreed that if kiln workers found that situation unsatisfactory, they had the option of giving up their traditional livelihood and becoming farmers in the reclaimed area.[42] By 1936 the brick and tile industry was in decline. Tiles piled up unsold because farmers in the depressed rural economy could not afford to buy them. The relatively poor quality of the clay made them unsuitable for foreign markets.[43] Making molds and firing the kilns, moreover, were handicapped by the centuries-long deforestation of some of the mountains around the lake.[44] The reclamation of Upper Xiang Lake was simply another blow to the industry's fortunes. The government's policy to the plight of the kiln households in the 1930s contrasts vividly with the benevolent paternalism of the Qing government in the 1770s, when great concern was shown lest such households be displaced from lake land.

During the 1930s the central government at Nanjing saw many of its reconstruction blueprints yellow with age, unrealized largely due to a lack of money, motivation, and realistic planning, ineffective administration, continuing domestic turmoil (fighting with Communists and "residual" warlords), and a new, virulent Japanese imperialism. Although reclamation proceeded apace at Xiang Lake, time showed that some of the central government problems, specifically poor planning and ineffective administration, also undercut this project. In spite of the surveyors' claim in 1927 that irrigation would not be harmed, later commentators reported that the reclaimers had not even considered the problem of irrigation for either the townships or the newly reclaimed land. As a result, a severe drought in 1934 spawned almost 300,000 refugees, a tragedy that adequate lakewater could have ameliorated. To irrigate the reclaimed land, several small-scale mechanical pumps were installed at Wen Family Dike to lift river water into the lake area.[45]

Even when irrigation water in the ponds surrounding the reclaimed land was sufficient, the reclaimers had not leveled the land, making it difficult to attain uniformity of irrigation. The vaunted easy transportation advantage of the area initially proved an empty promise. Roads were poorly built and indistinct in places; no bridges linked the reclaimed area with the original lakeshore, as at Zhang Village at Green Mountain. Since the reclaimed area's interior river was not dredged until 1933, transportation was inconvenient and sending produce to markets was especially difficult.[46] In sum, in spite of the superficial detail of the proposed schedule, there was no effective plan for or administration of the development project.

These problems compounded the economic difficulties faced by both farmers involved in the experimental farm and tenants of the province-owned reclaimed land. In January 1932 Zhejiang University initiated farm credit cooperatives at various villages, including Ding Mountain, Zhang Village at Green Mountain, and East Wang Village. Though relatively successful in the short run, the credit cooperatives were, on the whole, ephemeral. Production and consumer cooperatives had an even more difficult time. A fish-raising cooperative met immediate difficulties because farmers lacked experience with cooperatives and because outsiders stole fish from the cooperative's stock. A Ding Mountain Village food cooperative established in 1935 was also shortlived and ineffective in dealing with the rising price of rice.[47] An attempt to foster economic connections between people in villages, cooperatives joined the important impetus of self-interest to the traditional tie of native place. In a period of less economic exigency, the cooperatives may have had a chance at alleviating local problems. But widespread depression was compounded by the slipshod administration and mismanagement of the reclaimed area and made their success unlikely.

By the mid-1930s Xiang Lake could no longer be used effectively for irrigation. Its brick and tile industry was crippled. Its abundance of water-shield plants had disappeared; people now caught poisoned fish instead of the succulent catches of old. Buildings, statues, and inscriptions from its historical sites had been dismantled and stolen. The world of natural beauty, vitality, and productivity, the site of the centuries-long struggle between encroachers and lake protectors, had been overrun without restraint by predators and scavengers.

Ends and Means

In the destruction of the lake as it had existed, the traditional Chinese emphasis on means as opposed to ends also seemed to be disappearing. Much of the Chinese ethical code dealt with means, specifically the forms, procedures, and relationships conducive to the generally accepted end of social harmony. The Confucian bonds and various other social "connections" were relationships through which ends were gained. Education was the means by which official and elevated social status was achieved. Following the forms and procedures and abiding by the relationships were signs of the ethical and cultivated man. The emphasis was so strong that often forms, procedures, and relationships seemed to become the essence rather than the medium by which the essential was reached. There had always, of course, been "goal-oriented" people; the lake had seen more than its share of them: Sun Xuesi, Sun Kaichen, and Huang Yuanshou, to name the most egregious. But even they had depended on relationships, cultivating connections to officials and to local leaders and followers and building personal reputations as men concerned about others in the locality (in arenas outside of water conservancy).

In the late 1920s and the 1930s, official predators and their hangers-on seemed to have cast aside the traditional ethical emphasis on means. The goal or end of strengthening the state (and often enriching personal coffers in the process) justified destroying the livelihood of those who still depended on lakewater, undermining the traditional brick and tile industry, and dismantling historical sites. Relationships, forms, and local connections were dispensed with as government used its arbitrary power with seemingly little concern for the ramifications of its actions among local residents. In 1929, for example, the local government set out to destroy the relationship of dikes that had made irrigation in Yiqiao Township possible since the eighteenth century. The Lian sluicegate and Hengzhu Dike had, during the irrigation season, prevented merchant access further into the interior, but they were necessary for irrigating the township. Now they were to be changed to allow continual merchant access directly to the south gate of the county seat. Government representatives and local merchants were to make the decisions; farmers affected by them seemed inconsequential.[48]

Throughout the centuries Xiang Lake had been not only a source of irrigation water, fish, water plants, and mud; as attested by the body of lake poetry, it had also been a place of natural beauty to which men and women might retire for inspiration, solace, or simple pleasure. It had been the haunt of the living and the permanent resting place for many dead. All this also began to be changed. A Shanghai newspaper reported in June 1930, for example, that a cemetery on the western mountains of the lake had been moved to make room for additional housing for the agricultural experiment farm. The reporter commented that in the future the lake would lack its former beauty.[49]

The planned reconstruction of the natural beauty of Xiang Lake as it had existed in the past was best set forth by one Gong Ying in an article written in 1936. These are only more blueprints, but they jar in light of the lake's ancient natural beauty. In thrall to the West, like many of his contemporaries, Gong began his description in obeisance to a French economist who wrote of the importance of gardens and open spaces for the health of urban dwellers. Noting that lakes were natural gardens, he suggested, making clear his understanding of means and ends, that "if we establish famous places of scenic resort [around the lake], then we can attract tourists; we can subsequently use the money gained in tourism to uplift our commerce and industry."[50] Transportation in the form of a system of roads around the lake with connections to scenic spots in the mountains was a prerequisite—the "key to unlocking the treasure chest of scenic spots." The traveler should not, he averred, have the feeling of being alone.

In a section called "Improving the Scenery and Embellishing Its Method of Presentation," Gong wrote, "Natural jade is not a polished stone or a finished instrument. Even though natural scenery is beautiful, if there is no human to beautify it, then it's difficult to appreciate its great value. If there is an area covered by thorns or a place where miscellaneous grasses are growing, we must clear them away to protect the purity and beauty of the view."[51] To improve and embellish the scenery of Xiang Lake, Gong called specifically for adding hotels, restaurants, parking lots, boat docks, soccer fields, basketball courts, skating rinks, swimming pools, pagodas, arbors, a zoo, a museum, botanical gardens, and signs for directing the moneyed tourist to places of scenic interest. How such additions would "protect the purity and beauty" of the lake is not explained; Gong, like many others at the time, had mastered the art of the Honest and Upright Company. In Gong's plan, the simple natural beauties of the lake on which scholar-gentry throughout eight centuries had rhapsodized in elegant poetry would be swallowed up in reconstructive, pragmatic commercialism. Other than the destruction of the irrigative capacity of the lake, this plan for reconstructing Xiang Lake's scenery best symbolizes the disdain for the past and for "means" as measured against "ends."

The Xiang Lake Normal School

Just as Gong's ideas about the development of tourism derived from Western writers, the inspiration for the Xiang Lake Normal School also came from the teachings of a Western philosopher and teacher, John Dewey. The school was the project of the famous educator Tao Xingzhi, who had studied under Dewey at Columbia University from 1915 to 1917.[52] Promoting the unity of study and action (and thereby hearkening back to ideas of Ming dynasty philosopher Wang Yangming), Tao

Figure 9. Snow Scene: Xiang Lake Normal School at Yewu Mountain (c. 1929)

designed this normal high school as a training center for teachers for illiterate farmers in area villages and, in a broader sense, for tasks of local rural reconstruction. Tao's goal, like that of many contemporaries, was to construct a modern nation; the method he chose—education—was a traditional Chinese means for attaining political ends.[53] But the content of that education, traditional subjects plus vocational experience, differed sharply from the orthodox traditional Chinese view, which separated thinking from doing.

Tao's desire to establish another school similar to his Xiaozhuang Normal School near Nanjing had coincided with the reclamation of the lake for the Third National Sun Yat-sen University, headed by fellow educator Jiang Menglin, who was sympathetic to Tao's plans. The school was established at Yewu Mountain in the rebuilt Clouds of Xiang Temple, the main hall of which served as auditorium and cafeteria; its two wings contained classrooms, laboratories, a library, and an infirmary (figure 9). Two-thirds of a mile away on Ding Mountain (connected, in the beginning, by a student-built causeway) were constructed more classrooms, a dormitory, and a school for the children of Ding Mountain residents. Tao visited the site in early July 1928 during the Ding Mountain construction, which necessitated clearing thickets of pine, bamboo, and tall grasses as well as battling heat and swarms of mosquitoes.[54]

Classes began on October 1, 1928, with a few more than ten students. In the beginning students could set their own length of study, usually one semester or two. In 1931, however, the school abolished such flexibility and began a four-year system,[55] continuing to offer varied length "special courses" and night classes for farmers. The curriculum focused on science (physics, chemistry, biology, mathematics, agriculture) and education (principles of education, educational psychology, and elementary teaching materials and methods).[56]

The Xiang Lake Normal School became the area's most famous educational institution in history, far surpassing the significance of the Southern Path Academy on Pure Land Mountain of the Ming period. Experimental and practical, the education at the school thrust its students into the service of the larger society. Working on the thirty-five-acre school farm was required; whether the job was planting vegetables, raising fish, or harvesting rice, the goal was training for later contributions to agricultural labor and reform on the students' return home. As part of the emphasis on manual labor as well as "book learning," students also participated in building dikes, opening streams, digging wells, leveling land, repairing roads, and constructing various buildings.[57]

Before graduation, all students were required to teach at one of the elementary schools established by the Xiang Lake School in rural villages. Located at Ding Mountain, Zhang Village at Green Mountain, Stone Grotto Mountain, Chen Village, and outside of Wen Family Dike, these elementary schools were established at existing village structures—temples, shrines, or perhaps even traditional private schools.[58] Local farmers often responded to schools' establishment by nonlocal men with fear and confusion. A Xiang Lake teacher reported in 1929 that on a visit to three villages along the lake's northwest shore, most farmers were so frightened that they ran and hid.[59] Such a reaction points to the sorry record of outsiders in the whole reclamation experience and to inadequate planning by Xiang Lake School administrators. Their reception was much better when they arranged in advance to have a member of the local elite with "connections" to the village farmers organize a meeting. Local men with some measure of authority thus became brokers between outsiders and the local populace.

Because of the possibility that villagers might misunderstand the purpose of outside educators in their village, the Xiang Lake School policy required that the staff at each elementary school undertake home visits in villages to explain the purposes of the schools and to understand the home situations more clearly. To foster good relations with lake communities, the school became involved in village problems. No medical services and little medical knowledge were available in the rural villages. But when an epidemic of encephalitis broke out in Chen Village in 1931, Xiang Lake students accompanied the school physician to homes to treat the ill and to show those not yet infected how to defend themselves against the mosquito-borne disease.[60] Xiang Lake School also tried to carry forward Tao Xingzhi's "little teacher" movement wherein school-aged children became teachers to illiterate adults. The school leaders hoped that a modicum of basic instruction might entice those adults to come to school-sponsored night classes.[61]

The reformist ideals of Tao and the school at its founding are set forth clearly in the school poem, which effectively uses water in lake or bay as its central image:

> Undulating waves push each wave higher,
> The last led by others before it.
> People develop in mutual fashion:
> Those of the past provide instruction to inform life in the present.
> Education leads; the situations of life build.
> In the endeavor of leading, the later wave becomes even higher than the former.
> In the endeavor of building, those who live later build even better than their predecessors.
> We do not seek to be alike; we must excel over others:
> Look at the waters of Xiang Lake; look at the Zhejiang tidal bore![62]

Noteworthy in addition to the emphasis on education *and* life experience as keys to excellence is the gradual reformist theme of building on the past and on the accomplishments of others. Waves, whether higher or lower, are linked to other waves in the expanse of water. This fluid image expresses well the interconnectedness of progress in the present to the successes of the past.

For some leaders of the Xiang Lake School this image of evolutionary change would have been repugnant. In their view, the school, whose program reached out to rural communities of poor, illiterate farmers challenged by land developers, was a natural base for organizing political revolution. In 1929 and 1930 the head teacher of the Xiang Lake Normal

School was Yun Yiqun, simultaneously the head and acting secretary of the Xiaoshan county Communist committee. He organized the seven-member Xiang Lake Normal branch of the Communist party as well as the school's Communist Youth League. With another faculty member, Yun introduced Marxist thought to students and led discussions about the history of the Communist success in the Soviet Union. Yun also accompanied students to area villages to conduct social surveys, in the process speaking with and encouraging peasants about revolutionary ideas.[63] For Tao Xingzhi, the school was a vehicle for gradual, if wide-ranging, rural reconstruction; for Yun and his supporters, it became the means to a revolutionary end. It was the same split (gradual pragmatic change versus revolutionary transformation) that had put intellectuals all over China on separate political paths in the 1920s.

The revolutionary activity at the school and in nearby villages could not remain secret from the Nationalist party, which controlled the state. In November 1930 the provincial police bureau sent military police to arrest Communist party members and revolutionary students at the school and to effect changes in its organization.[64] With the arrests came the ouster of the principal and the expulsion of several nonparty students. The back of the county Communist party was broken; it would not be rebuilt until the Japanese invasion radically changed the political situation.

The destruction of the Communist party did not, however, end Nationalist party suspicions about the Xiang Lake Normal School. Any report of school policies or actions that could be tagged "progressive" brought intimidating threats from the government. In 1931 the Nationalist government ordered that radical or "progressive" books in the school library be sealed and put away. Faculty and students first hit on the idea of distributing the offending books among students so that the library itself would not contain politically questionable works. But a number of troops were dispatched from Wen Family Dike to search the entire school premises. This news reached students only a frantic few minutes before the troops arrived, but it was enough time for the books to be hastily gathered and thrown into a small boat, which was rowed to the deepest part of the lake. The books were dumped into the murky water, and the students avoided arrest and political retaliation. Added now to the anguished suicides the lake had accepted in previous centuries during chaotic political change were books on revolutionary change, sunk to their watery repository out of fear of an increasingly authoritarian, vengeful Nationalist government. In the years that followed, the government continued its pressure on the school, calling for Xiang Lake Normal School to be controlled by teachers guided solely by Nationalist party principles.[65]

Like the lake for which it was named, the school was to be reconstructed in the particular image of political powerholders who seemed increasingly distanced from local reality. Few of the local or provincial players in the drama of Xiang Lake could have guessed that within a few short years another of the periodic military scourges visited on the area through history would open once more the larger question of local ends—specifically, the disposition of Xiang Lake.

IX

VIEW

From Little Li Mountain, Late 1980s

CHAPTER

Transmutations, 1937–1986

The old heavenly bodhisattva is thoroughly cruel
To keep sending the rain
And stirring up the wind;
Mother A Da's clear eyes
Have become red with weeping.

The Qiantang river!
When it floods, people seek refuge,
But everywhere they eat bitterness.
Older brother is returned home, dead;
Younger brother returns home, but fate forbids him stay.
If I go now, out to the river,
I swear I won't rest until I kill the demon!

—Two Xiaoshan folk songs, collected 1981

VIEW: *From Little Li Mountain, Late 1980s*

Almost two centuries ago, in autumn 1797, Yu Shida climbed to the top of Li Mountain.[1] The vista included rice fields on two sides of the mountain, the lake to the north, and the river to the south. On this visit he met a friend, a lifelong resident of the area, who spoke of the locality's past:

> When I was young, I heard my grandfather talk about the river, about how the great force of water from the areas to the south smashed against the river dike, several times destroying it—calamities which even led some to say it was pointless to have a dike. From the Ming until now, the body of the dike has been reconstructed three times, but the amount of paddy land still collapsing into the river is great.

In the late 1980s the view south from the same mountain (now called Little Li Mountain) provided the still awesome sight of the powerful flowing river. Even on a day when no rain has fallen, one can see why local

Figure 10. Juncture of Fuchun, Puyang, and Qiantang rivers near Wen Dike; reclaimed area, left (1986)

farmers, fishermen, and tilemakers could ascribe the rage of flooding rivers engorged by storms to malevolent demons. From the southeast comes the Puyang River, rough and churning over a steep riverbed to its meeting with the Fuchun River. The two rivers form a great expanse of water less than half a mile southwest of the mountain (figure 10). There, near the town of Wen (Family) Dike, the wedded rivers become the Qiantang, at least a mile wide at that point (see map 13). In floods, this expanse becomes a massive raging torrent, and the consequent disasters of the past thirty years are a catalog of tragedy:[2]

June 1955. Typhoon. Water level at Linpu on the Puyang River reached an all-time high (10.53 meters—the usual depth is 3 to 4 meters). Dikes broke; 13,000 acres were flooded.

August 1956. Typhoon, the eye of which passed over the county. The grim statistics: the river dike broken in 166 places; 12 water gates washed out; 263 separate dike embankments destroyed; 19,744 homes destroyed; 49 killed; 291 injured.

September 1962. Typhoon. Wen Dike received over 10 inches of rain

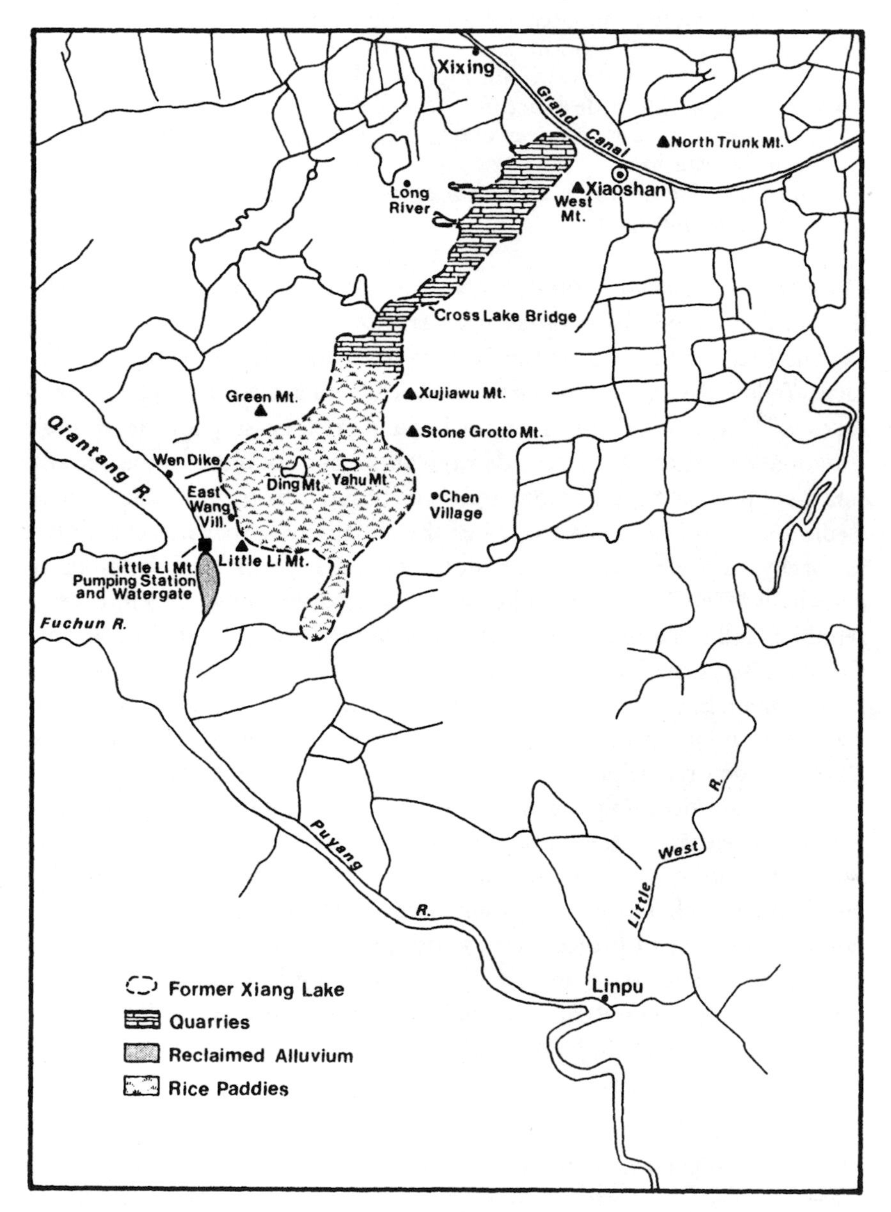

Map 13. Former Xiang Lake Region, 1986

with almost 9 in one day. Dikes were broken in 79 places; 66 villages were inundated; almost 60,000 acres were flooded.

September 1963. Typhoon. River water poured into broken dikes after 17 hours of heavy rain and high wind; 8 villages were inundated.

August 1974. Dike destruction after a storm led to the flooding of 39,000 acres and the destruction of 18 bridges, 1,315 homes, and 2 electric stations.

June 1984. Heavy rains broke dikes in 23 places, flooding 6,000 acres of cropland.

The demon has been the county's constant companion; throughout history Mother A Da's eyes have rarely been free of tears.

The frequent flooding of the Puyang River has led people to call it the Little Yellow River, a reference to the north China river denoted "China's sorrow."[3] Like the Yellow River, the Puyang carries vast quantities of silt, a situation exacerbated by the deforestation of mountain slopes to the south—a process begun before 1949 but brought to fruition during Mao Zedong's Great Leap Forward, when the remaining timber was cut for fuel for backyard steel smelters and building materials for mammoth commune mess halls.[4] The West River Dike had been repaired many times during the republic, and since 1949 the government of the People's Republic and the Communist party have strongly emphasized controlling this river. Its proper management affects not only the Xiang Lake region but all of Xiaoshan and much of Shaoxing county to the east. Flood control policies have included repairing the dike, applying stones along the dike's base, widening and dredging the river, and widening and dredging collateral streams. These have often necessitated large-scale construction efforts and the mobilization of from half a million to one million workers. The riverbed through Qiyan Mountain, excavated first in the famous mid-Ming water control project, was widened from 55 to 180 meters in the period from August 1973 to July 1976.[5] As no government since the Ming dynasty, the People's Republic has kept the demon continually in mind, not with the impossible goal of killing it but of keeping it in check.

Directly east of eighteen-meter-high Little Li Mountain was a newly planted rectangular area of land reclaimed from the river. The interplay of river currents and the lay of the riverbed had brought about the deposit of large amounts of silt. As county residents had done since the beginning of time, people began to reclaim it, more careful than at any time in history to dike it and maintain it vigilantly. From the mountain an assortment of

Figure 11. Little Li Mountain Watergate (1986)

vehicles was clearly visible on the gravel road that threaded its way atop the river dike: trucks, buses, makeshift tractors, bicycles—all vied for space with children walking from school, women carrying baskets, and farmers pulling loaded wheelbarrows and wagons. Between the mountain and the town of Wen Dike along the river was the Little Li Mountain watergate and pumping station. Established in August 1960, the pumping station was the county's first electric irrigation pumping station to produce more than two hundred kilowatts of power. The associated watergate provided a safety valve for the outside river at time of flood (figure 11).[6]

Just to the northwest of the mountain was East Wang Village, site of one of the credit cooperatives of the 1930s, now with a population of almost 1,100. People there still talked of the especially brutal devastation brought to this area by the Japanese invaders almost fifty years earlier.[7] To the northeast the yellow-green of ripening rice stretched from Green Mountain on the left to Stone Grotto Mountain on the right—the old site of Upper Xiang Lake. In this expanse of rice were islands of small villages, ribbons of crisscrossing canals, and a row of utility poles running from Little Li Mountain toward the former site of Lower Xiang Lake, carrying electricity to power pumps, lights, and small machines. Often paralleling the path of the electrical wires and running past Yewu and Ding mountains, little more than small hills, was a dirt road headed toward the old Cross Lake Bridge. In the continuing mongrelization of traditional names, Yewu (Thrown into the Crow [river])—a name charged with despair—had in the 1920s become Yehu (Thrown into the Lake)—a neutral and innocuous name, given the geographical surroundings; in the 1980s it had been further changed to Yahu ("duck lake"), which, locals explained with pride, took its name from the area's great success in raising ducks. The original name's sense of tragedy with its obscure allusion to the ancient Chinese past had been replaced by one that bespoke a present-minded rural productivity.

If one's eyes followed the dirt road to former Lower Xiang Lake, through the haze one could make out the fifty-foot-tall kiln smokestacks of the brick and tile industry. It was the haze, however, that all too quickly became the annoying focus, not the haze of hot summer afternoons that would dissipate in the cool of a Xiang Lake autumn nor the smoke of domestic fires, tranquil tokens of evening rest and familial sharing after labor in the fields. This noxious haze obscured the outline of Green Mountain two kilometers away and Stone Grotto three kilometers distant. It was air pollution from a complex of chemical factories built at the town of Puyan five kilometers from Little Li Mountain and from a fertilizer factory, much nearer, just outside Wen Dike.[8]

If Xiaoshan residents, under continual official vigilance, had begun for the first time in history to control more effectively the demon in the Puyang and Qiantang rivers, their general disregard of their natural environment and resources seemed flagrant. Brick- and tile-makers early in the century might pay scant attention to patterns of mud-dredging, a lack of care that might play havoc with lake boating but would affect few people. And an occasional "bad" fish might poison Xiang Lake denizens (as in the case of Sun Quwen in 1929). But, after the 1960s, in the Xiang Lake area the problem of chemically polluted water, killing fish and causing widespread human illness, grew serious.[9] Shadowed by the specter of poisoned air and contaminated water, one could descend Little Li Mountain pondering whether—given all Xiaoshan's tragic invasions—an even more malevolent demon had invaded the scene with the uncontrolled pollution of modern technology.

CHAPTER: *Transmutations, 1937–1986*

The years 1937 to 1945 brought to the Xiang Lake region a Japanese military invasion, a nightmare of fiery savagery reminiscent of the Taiping scourge of eighty years earlier. The conflict had seemed almost inevitable. After six years of acceding to Japanese demands for territory and privileges within China, Chiang Kai-shek was forced by his own bizarre kidnapping in 1936 and by a rising tide of national sentiment to resist any further Japanese threats. Japan's determination to adhere adamantly to its position produced a situation where a minor incident near Beijing in early July 1937 escalated into full-scale war. Fighting spread to the south in late summer, and Shanghai fell to the invaders after a three-month battle.

In July the Xiaoshan county government commandeered county inhabitants to construct defensive positions along almost two miles of the Qiantang River bank opposite Hangzhou.[1] Not only had the spirit of Li Di been ineffective in protecting the county from a Hangzhou-based offensive during the last attack (of the Manchus in the 1640s), but in 1930 the temple bearing Li Di's name on North Trunk Mountain had been destroyed.[2] The county might as well have depended on Li Di, however, for all the good earthwork defenses did: the first attacks came by air. On November 30, thirty Japanese planes attacked the county seat, killing 200 and injuring many. During the next six and a half months, Japanese bombers struck the county sixty-eight times, killing 604, injuring more than 2,200, and destroying almost 4,500 homes.[3] Major county institutions began to move into the interior. The county offices, destroyed in the initial bombing, were relocated first to the east in the town of Qianqing and in March 1938 to Heshang, south of the Puyang River.[4] The Xiang Lake Normal School, under the principalship of Jin Haiguan, a former student of Tao Xingzhi, left Xiang Lake, moving south first to Yiwu county and later far into the interior to Songyang county.[5] On December 23, 1937, to stop what seemed an imminent Japanese advance from Hangzhou to eastern Zhejiang, the Nationalist government blew up the bridge across the Qiantang River, completed only two months earlier.[6]

The bombing campaigns on the county continued; one of the most vicious killed several hundred people in February 1939.[7] But the crucial blow was the infantry invasion across the river from Hangzhou.[8] On the evening of January 21, 1940, amid a snowstorm, Japanese forces crossed the river and secured a base, moving the next day to capture the former county seat. The invaders came into the Xiang Lake area by two routes. From Long River they moved toward Wen Family Dike, and from the county seat they entered along the mountain range to Stone Grotto Mountain and thence to Yewu Mountain. Some troops then moved from Chen Village on the southeast shore of the lake to the town of Linpu.

Japanese troops and local forces fought a bloody battle in the paddies between Stone Grotto and Yewu Mountains, an area that less than two decades earlier had been covered by lake water. This region and its villages were devastated. All the facilities of the Xiang Lake Normal School, the agricultural experiment farm, and the agricultural extension office were burned to the ground along with any remaining wood on the mountains, crops in the fields, even grasses along the roads. The work of education and reconstruction begun by the Xiang Lake Normal School turned literally to ashes in the burning of most of the elementary schools opened idealistically a decade before. But, almost as if to emphasize how short the goals of the educators had fallen among the local populace, at Xujiawu on the east side of the lake, local people themselves, seeing the flames of the burning lake district, set fire to and destroyed their own school facilities.

Modern warfare brought death both by fire from the heavens and earthbound soldiers and by something yet more sinister. Refugees in earlier lake military tragedies could flee to safety, hiding in rough mountain terrain or even in water. But now the very air brought death: on August 14, 1941, Japanese forces stationed at the foot of Stone Grotto Mountain launched a late afternoon attack on the town of Yiqiao. Among the bombs dropped were two of poison gas. More than sixty people in this old center of the Han lineage died of chemical poisoning. Each year until the war's end, fighting ravaged the county as the Japanese continued to raid the towns of Linpu and Heshang, the relocated county seat, which remained outside the Japanese perimeter of control.

At war's end, Xiaoshan and the lake region had seen the most serious destruction of all of eastern Zhejiang. Almost forty thousand homes were destroyed, and, according to county population statistics, at the end of 1945 (after most refugees had returned) the population had shrunk by almost 79,000 from December 1939 (a drop of 15 percent).[9] Institutions as well as refugees returned once again to Xiaoshan. The Xiang Lake Normal School, under Jin's leadership, returned from exile and discarded its wartime practical-training curriculum of mountain climbing, marksmanship, and bomb throwing.[10] It relocated permanently in the county seat, abandoning the opportunity to foster education directly in the rural villages but not relinquishing its progressive views.

An immense task of reconstruction awaited those who returned to the county. All the commercial prosperity of the early twentieth century and the tentative moves toward industrialization had been gutted. The fishing, salt, and tobacco industries, all prosperous before the war, were destroyed.[11] Sericulture and tea production were brought to the edge of extinction because farmers had been forced to cut mulberry and tea plants for fuel.[12] The Japanese had dismantled the three modern textile mills for machine parts for their campaign.[13] The profitable lace industry and the struggling paper industry were both closed by the Japanese.[14] Whereas before the war most houses in the county seat were two-story wooden structures with tile roofs, the great postwar poverty spawned thatched huts to replace the thousands of bombed houses.[15]

Xiang Lake in the Postwar Years

The war with Japan not only destroyed Xiang Lake's reclamation facilities, ended its reconstruction plans, and further impoverished the area but also again opened the lake's future to question. Kong Xuexiong, one of the two principal compilers of the 1927 report that first recommended reclaiming the lake and longtime Nationalist party leader, came forward in 1946 with a proposal to establish a private company, the Xiaoshan Reconstruction Company, to "reconstruct" the Xiang Lake area.[16] He called specifically for rebuilding farm homes destroyed in the war, repairing public roads, dredging boat channels, fostering the recovery of fish raising and brickmaking, developing better transportation and communication channels, and establishing small factories. Kong's plan is noteworthy more for what it omits than for what it includes: nowhere is mentioned irrigation, even for reclaimed lands. The agriculture extension agent, Han Yanmen, also proposed in 1946 that Lower Xiang Lake remain an effective reservoir for irrigation (and a base of mud-dredging) while Upper Xiang Lake be reclaimed. Reports in 1947 stated that 10,000 mu (about 1,600 acres) of Xiang Lake still existed. Kong's proposal and Han's suggestion did not meet widespread approbation. Kong's plan seemed a déjà vu of the early republican attempts to establish private development companies, a certain red flag to area residents and other interested parties in a context where the legal status of the lake was still undetermined.

In 1946 and 1947, in fact, bitter contention erupted once again over the meaning of "public" in relation to the lake.[17] It was generally agreed that the alluvial (and now mostly reclaimed) land was public land—but *which* public? Was it controlled by the province? By the county? Or, as local residents would have it, by the areas touching on former lake land? The battle that had raged for half a century obviously had not ended. Out of the contention came the formation in 1946 of the Xiang Lake Rural Reconstruction Association, whose declared purpose was to lift the lake region out of economic depression. Joining in the establishment of the association were the region's respected elders (*fulao*) along with representatives from Zhejiang University and the provincial government's Reconstruction Bureau.

The association is significant; as commentators noted, it was the first time since 1926 that the government recognized that local leaders and residents should play a role in the disposition of Xiang Lake. The provincial government had backed away from its policy of bureaucratic control of the situation: the organization's principles stated that Xiang Lake belonged to the people around the lake and that they should reap the benefits of its resources. Remaining lake land was to be properly dredged so that the lake could function effectively as a reservoir and flood-control basin. Decision-making authority should be lodged in the lake area, not in the Hangzhou government, in the offices of Zhejiang University, or even in county bureaus in Xiaoshan. To achieve regional prosperity, local energy and enthusiasm would be directed and channeled by the provincial assistance and advice of Zhejiang University for educational reform and of the Reconstruction Bureau for agricultural expansion.

The association remains one of the "what might have beens" of the lake's history. Its chance of developing into a significant organization depended on enough time and sufficient political support to realize progress, on effective and conscientious local administration, and on economic stability. It enjoyed none of these preconditions. In little more than two years the Nationalist regime collapsed. In that period, moreover, the county board (*canyihui*), in the continuing struggle over the lake's status, voted that the ownership and management of the lake area belonged to the county.[18] Renters of land would have to negotiate rental agreements with a county board that administered public property and funds; a time limit for such negotiation was imposed, after which land could be rented to someone else.

The political and economic situation in the county was disintegrating with such frightening speed from 1947 to 1949 that success of the association would have been doubtful even had the county board not attacked. From the close of the war with Japan (August 1945) to the Communist victory (May 1949), the county had three different magistrates with an average tenure of fifteen months, insufficient continuity for effective leadership. In mid-1948 the county was still only talking about reconstruction rather than actively undertaking it.

When asked about the agricultural extension office's plans to bring agricultural university graduates to help reform county agriculture, Magistrate Hua Guomo was quoted as saying, "I can say this with certainty: if we can go according to plans, within two years we can bring remarkable progress to agricultural organization, economy, and production—indeed, farm life, in general."[19] As he was mouthing the proper words, Magistrate Hua, graduate of Chiang Kai-shek's Whampoa military academy, was playing the Upright and Righteous Company game. In early 1949 Hua, who also headed the county tax office, conspired with branch tax heads around the county, the head of the county granary system, and key rice merchants to defraud the public.[20] Because the county grew insufficient rice to feed its people, the magistrate, under the guise of raising funds to purchase rice outside the county, made available for sale "emergency coupons." He claimed that the coupons could be used to exchange for rice or even as payment for the land tax. Before the coupons went on sale, however, the magistrate had all the rice in the county granaries secretly removed to the various rice merchants' stores, dividing the money from the coupon sales among the conspirators. When people came to exchange the coupons for rice at the county granaries, they found the granaries empty and the coupons worthless. Meanwhile the merchants charged exorbitant prices for rice they sold. Hua was able to flee the county when the scheme became known, though the subordinates of the conspiracy were tried and punished in early 1951.

Such blatant corruption and disregard for people's welfare among county officials obviously did not provide the kind of leadership for reconstruction projects to flourish. As if such leadership were not bad enough, the leader of the Nationalist party in Xiaoshan, one Yu Xie, hatched a scheme to use pumps to remove water from the lake and reclaim more of the land on the basis of his own authority. His actions drew sharp protest, but it was the Communist victory that ended these self-aggrandizing plans.[21]

Even if Hua, his predecessors, and the leaders of the party had exercised conscientious and competent leadership, the economic disintegration of those years would have immobilized further action. The disastrous inflation that gutted any remaining support for Chiang Kai-shek's government in the late 1940s and prepared the way for the Communist victory can be illustrated most succinctly with a few mundane statistics. On April 23, 1949, two weeks before Xiaoshan was captured by Communist troops, a picul of rice (about 133 pounds) in the county cost more than 8,640,000,000 yuan, 785,400,000 times more than in 1937. If one figures that a picul on average contains about 3.5 million grains of rice, the price of each *grain* of rice was about 2,500 yuan.[22]

The Xiang Lake Rural Reconstruction Association, like so many other hopeful enterprises during the republican period, aborted. Even so, more plans for reconstruction were continually made. The second issue of a county Nationalist party reform journal devoted to the urgency of land reform and rural change (*Xiaoshan Land Administration*) appeared on January 1, 1949. Though it is filled with articles proposing specific programs and castigating Communist figures and policies, for our purposes it is significant because it contains the last recorded paean to a dying Xiang Lake—in effect, an elegy. As such, it is the last of the genre of lake poetry, incorporating many previous themes and images. The author, Tang Boyin, about whom there is no information, gave it the unabashedly ingenuous title "I Love Xiang Lake."[23]

> If West Lake is a city woman in full and fancy dress,
> Then Xiang Lake is a plain village lass with cotton skirt.

Xiang Lake's surroundings lack the beautiful pavilions and tall foreign buildings of Hangzhou.
But it has many famous tombs and the old colors and smells of thatched huts and bamboo.

When you walk around the dikes and look far away at Zhang Village at Green Mountain in the early morning,
The white clouds, pushed up like cotton balls, gather at the mountain top.
The mountain birds fly in flocks out of the forest:
Their cries make you look up, as they fly you know not where.
From the village at the water's edge there is a small boat
In which sit one or two disheveled, coarsely dressed country girls
Lightly, lightly dipping their oars into the water, moving slowly along.
Sometimes one or two voices singing a mountain song enter your ears.
It all seems like a picture.

Xiang Lake's four seasons have four different faces.

Although the spring has no peach blossom reds or willow green,
The luxuriant leaves of the forest trees surround thatched houses and village huts.
Far in the distance a great shout welcomes guests
Coming on boats from town to sweep the graves of their ancestors,
Sailing past people carrying baskets from the town
Coming to see the beauties of spring at the lake.

When summer comes, the bright sun reflects in the water.
At noon, except for a few small fishing boats
There is no trace of people in the heat.
Small children hide beneath tall trees, playing blind man's buff.
In houses women sit, spinning cloth:
The hum of their machines adds to the poetic spirit of the lake.

The autumn is Xiang Lake's most lonely season.
The boats of spring and the leaves of summer have gone.
The water is deserted.
A single water buffalo lies on the earth, an image of loneliness,
A loneliness which brings pensive thoughts of the past.

Winter brings the white snow,
Dressing Xiang Lake with powdery clothing—like carved jade.
The north wind blows cold on the face of the lake;
No one walks on the dikes.
The farmers are almost all in their homes, spinning, or crowding around the stove.
Very few townspeople go to see the beautiful winter face of the lake.
But those who do find Xiang Lake at its most beautiful.

Xiang Lake—a plain village lass with cotton skirt.
I love Xiang Lake.

The Critical Decision

At 2:00 P.M. on May 5, 1949, the Jin(hua)-Xiao(shan) military detachment of the Communist People's Liberation Army entered the Xiaoshan county seat and received the surrender of the Nationalist government: "liberation" had come.[24] The Jin-Xiao detachment had been established in the summer of 1941 as an anti-Japanese fighting force of local partisans along the Zhejiang-Jiangxi Railroad, which ran south through Xiaoshan and Jinhua counties.[25] Nominally dispersed in September 1945, the detachment was active by 1948 in the anti-Nationalist campaign.

Under the forceful leadership of Jin Haiguan the Xiang Lake Normal School was an institutional feeder of men to the detachment. In May 1947 the Xiang Lake Normal branch headquarters of the Chinese Communist party was formed; that same month over six hundred of its teachers and students demonstrated in surrounding towns, opposing hunger, war, and government repression and reportedly making an impact on elementary school students and area farmers, among others. The school's active political role made it, as in the 1930s, an object of Nationalist repression. In October 1948 the school party branch was outlawed; as a result, many of the school's students enrolled in the Jin-Xiao detachment. During the last half of 1948 the detachment made damaging attacks on police posts in major towns; and in 1949 it began to seize the towns themselves. When the detachment liberated Xiaoshan, fully one third of Xiang Lake Normal School's teachers and students were members of the People's Liberation Army.[26]

As Xiang Lake awaited its final disposition, the early 1950s were filled with rapid, sometimes violent change for the people of the region. Late 1949 and 1950 were marked by frequent attacks on Communist functionaries by opponents of the new regime—labeled "local bandits" by the Communist sources. Such attacks were often followed by executions of the so-called disturbers of the peace.[27] The centerpiece of the Communist revolution—land reform—was instituted in November 1950, begun with two trial districts: Long River, the old base of the Lai lineage, west of Xiang Lake, and Linpu on the Puyang River. The first stage of the process, "land to the tiller," was completed by July 1951.[28]

Amid the launching of land reform came some of the county's "show trials" of enemies of the people, individuals described in such unbelievably monstrous terms that we can take them mainly as symbolic incarnations of evil and therefore official targets for vengeance. In January 1951 county seat resident and member of one of its famous lineages Chen Xiangfan (literally Chen "Xiang [lake] sail"), local leader since 1929, collaborator with the Japanese, and alleged tyrannical landlord charged with murder (of over 140 people), theft, and assorted other crimes against the people, was executed after being found guilty by the Xiaoshan People's Court.[29] More Japanese collaborators and those found guilty in the preliberation rice coupon fraud were executed in March. By that time, action against opponents of the regime was facilitated by the supercharged atmosphere produced by China's participation in the Korean War against the United States. The Resist America, Aid Korea campaign was especially vigorous in Xiaoshan; increased production and "contributions" reportedly totaled enough to purchase six bombers for use against the United States.[30] Certainly the numbers of young men from the county who were killed, let alone wounded, in Korea would indicate that many who contributed probably had connections (either familial or native place) with casualties. The 101 men killed included 4 from the Lai lineage of Long River and 1 each of the Han, Ni, and Kong lineages of Yiqiao.[31] Waging both domestic violence and foreign war, authorities had little time to deal with the reconstruction of water control facilities.

It was not until February 1952 that the county's great model of postliberation waterworks construction was undertaken: widening the Puyang River south of Linpu. By May a total of 840,000 workers from six neighboring districts had participated in the project. A much less spectacular development, however, in the spring of 1955 spelled the end for Xiang Lake as it had existed: at Long River, with the use of national capital, the county's first diesel irrigation pump began operation. For centuries the treadle water pump (alleged to have been invented by a native of Xiaoshan county) had been used to lift water from canals and ditches to paddies.[32] Completely dependent on human leg work, this machine was impracticable for large-scale irrigation. In the 1920s the Xiang Lake agriculture experiment farm imported a few six- to eight-horsepower diesel pumps from the United States, but they were useful only on a small-scale basis.[33] The pump at Long River was the first in the county with the capacity to irrigate large acreages.

That year, 1955, the decision was made to reclaim all the lake, which still covered 10,000 mu (1,600 acres, or 2½ square miles) at depths of five meters or less. Ironically we know as little about the decision to destroy the lake as about the decision eight and a half centuries earlier to create it; records of the decision, if they exist, are unavailable. Discussions in autumn 1986 about the disposition of the lake revealed, however, the decision must have been controversial.[34] Dong Mingzhi, chief engineer of the county water control office, argued simply that once the technology of irrigation through large-scale pumping had been attained, the lake lost its value. In earlier times, the tidal water from Hangzhou Bay had sometimes traveled as far upstream as Linpu, making the water in the Puyang and Qiantang rivers too brackish for irrigation, even had the technology existed. Beginning, however, in the eighteenth century, the Qiantang had turned sharply north, depositing its huge load of silt in new patterns to create thousands of acres of reclaimable alluvium. Whereas before the eighteenth century the town of Wen Family Dike was about twenty kilometers from the bay, in 1955 it was about seventy kilometers. Water, therefore, could be pumped from the river without fear of salinity.[35]

Yang Jun, teacher at Xiang Lake Normal School, did not dispute the role played by technology, but he argued, noting how much of the lake still existed when he arrived to teach in 1948, that the value of the lake for irrigation, though diminished, was by no means lost. He pointed out that the lake decision was made in 1955, fully five years *before* electric pump technology came to the county in the form of the first electric pumping station (Little Li Mountain Pumping Station), which was established to irrigate the reclaimed lands of Xiang Lake and some surrounding acreage. Implied in Yang's arguments was a suggestion that the lake could have been maintained and pumps installed to make it serviceable for irrigation. The discussion of the three-decade-old decision was polite; the disagreements were muted. Yet one could sense the disdain Dong felt for the nonspecialist Yang and the obvious regret Yang felt over the lake's demise. In 1955, the counterarguments to Yang's position would have been that the lake was an exhaustible source of water but, more importantly, that reclaiming the lake would answer the need for more land at a time when modern irrigation technology obviated the need for a lake. Small canals in the lake area were to be maintained only as channels for boat travel.

Thus, the last phase of the reclamation of Xiang Lake began in 1955 as a number of work and residential units—the Xiang Lake agricultural experiment farm, the towns of Xixing and Stone Grotto Mountain, and

Figure 12. Xiang Lake Canal (1986)

rural villages around the lake—turned water into earth. Within a decade, over 6,000 mu of new land were producing rice, wheat, and rape. In addition, brick- and tile-makers within the same period took over 1,000 mu to dredge for clay. An investigation in 1966 showed the lake area to be only 3,040 mu (about 490 acres, or 3/4 square mile). In the late 1960s seven more brick and tile factories took additional land; the Hangzhou Cogwheel Factory, completed in 1965 and located west of Pure Land Mountain, took former lake land; the pools and tanks of the Xiang Lake fish hatchery, the county's chief fish-raising facility, established in 1976, was located in lake land to the south of the Cross Lake Bridge.[36]

In 1986, apart from public roads built on former lake dikes, transportation and drainage canals, and a few diked ponds—a total of about 1,400 mu (about 224 acres)—all of Xiang Lake had been reclaimed as cropland or was used for brick-making. Yet despite the transmutation to earth, officials seemed almost reluctant to admit that the lake was gone, perhaps another indication of a sense of loss of this county-identifying historical legacy. On an official map of the county seat and its environs, produced in 1986, the main forty-meter-wide canal running through what was formerly the lake is denoted as Xiang Lake (figure 12)! It is almost as if the name itself could transmute the "canal" into "lake"; it is the same legerdemain and potency of nomenclature that has so dominated social and political life in the lake region during the twentieth century.

Xiang Lake: A Postmortem

Even before the lake's destruction, many of its natural aspects and historical surroundings had disappeared. As the water became shallow, many of the lake's prized fish died out. The water-shield plants, largely uprooted in the Taiping aftermath, were destroyed. The once plentiful wild ducks had flown to other lakes; the gulls disappeared as the Qiantang alluvial deposits drove the ocean many miles farther away. Of the once abundant freshwater crabs, few were left. The surrounding mountains and the former island mountains were deforested of pine, bamboo, the arbutus, the tallow, the *wutong*: the demands of the brick industry, Japanese scorched-earth warfare, and the insatiable demands of the Great Leap Forward (1958–1959) had worked their destruction.

All the pavilions, shrines, temples, springs, and historical markers have been swept away through the cruelties of nature and man. Only rubble remains at certain spots on the King of Yue Garrison, Yewu Mountain, and Stone Grotto Mountain. One View Pavilion on Stone Grotto was apparently the last major landmark to be destroyed after 1947, presumably during the Communist period.[37] The site of Mao Qiling's home in the county seat is the county government and Communist party headquarters. Longxing Temple, reconstructed in 1887 at the base of Pure Land Mountain, by the 1980s served as a government granary.[38] One other change at West Mountain (now the name for the series of mountains running from Pure Land Mountain to Stone Grotto Mountain) is noteworthy. In the late 1750s this chain of mountains was the site of quarrying efforts that were beaten back by local leaders under the charge of disruption of natural geomantic forces. By the 1980s this very mountain had been substantially quarried from underneath: an internal tunnel connected two parts of the Hangzhou Cogwheel Factory (see map 14), but most of it was hollowed out to form the Xiaoshan county People's Air Raid Shelter with a large underground theater and the people's "defense hostel."[39] Geomantic concerns had been replaced by the demands of a modern state and of defense against yet another military attack sometime in the future.

It was indeed a changed world, but one that inevitably led back to the

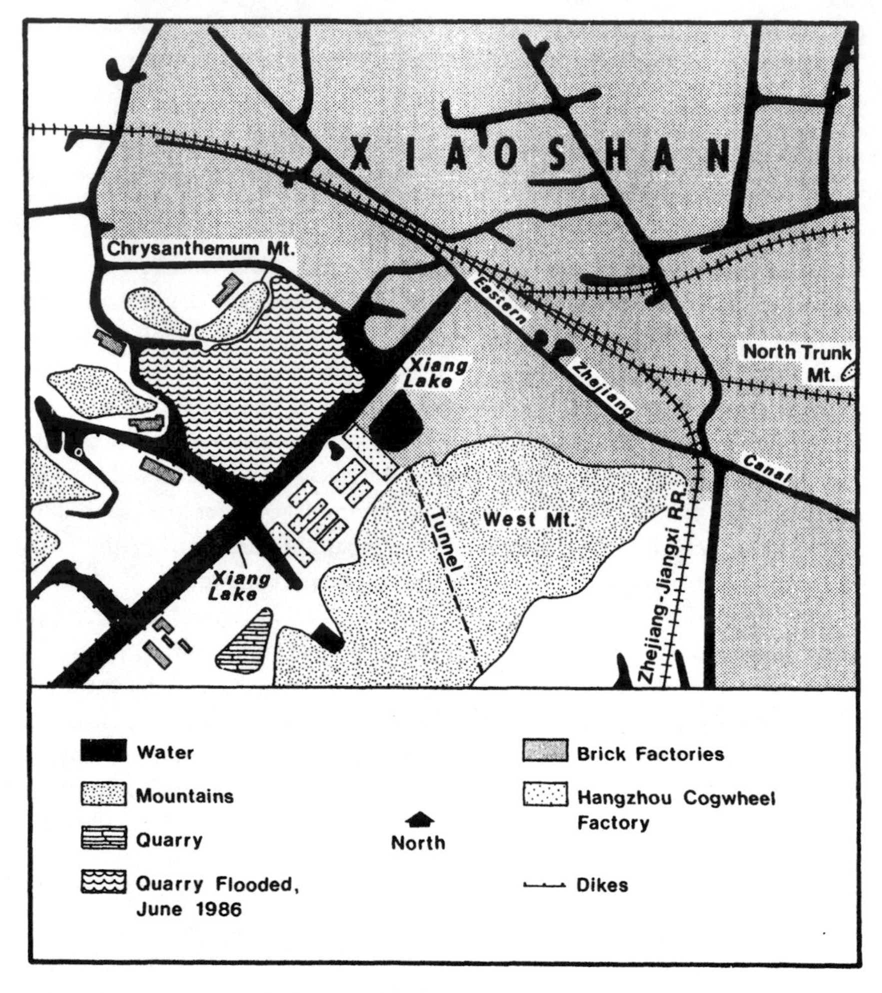

Map 14. Lower Xiang Lake Environs

theme of many early poets in Xiang Lake's history—time's obliteration of the past. The images of repeated tragic invasion and death in the poem of Sun Xuesi, the builder of the Cross Lake Bridge and major contributor to the lake's eventual destruction, seem a mournful epitaph to the lake and its region.

> At night it is as if there were rank upon rank of soldiers engaged in ghostly battles.
>
> The moon is black over a thousand peaks.

To broach this theme, of course, is to bring to mind that the death of Xiang Lake came fundamentally not because of the modern technology of electrical pumps but as the conclusion of an eight-and-a-half-century drama staged by nature and man. Reclaimers and irrigators struggled in the main story line almost continually; in the last acts, officials and a changed political and technological context gave the reclaimers the chief role. But the action was punctuated, halted, and redirected, on the one hand, by the storms of nature, the flood and silt of rivers, the pounding surf of the bay and, on the other, by the bloody brutality of human beings attacking and destroying other human beings in periodic invasions, each attack more deadly than the last. Shaping the action through their hold on the actors were the power of traditional moral values and the acute consciousness of the past in the living of the present. The remarkable continuity and managing role of the leaders of area lineages gave the drama a social framework. A host of minor themes were elucidated by the actions of the chief characters and supporting cast: among them, the dominant motif of subordination to authority, the changing styles and methods of public stewardship in managing the reservoir, the crucial role of local officials and their changing relationships with local leaders, and the self-conscious articulation and debate beginning in the eighteenth century about the meaning of "public."

But perhaps the most surprising aspect of the drama was that the chief character and setting—Xiang Lake—endured as long as it did. While every other lake in the area (except for West Lake) was reclaimed during the Song or Ming dynasty, Xiang Lake remained.[40] Such life must be attributed in large part to the long list of lake heroes who sacrificed to perpetuate it: Yang Shi, Zhao Shanqi, Gu Zhong, and Guo Yuanming of the Song; Wei Ji, He Shunbin, and He Jing of the Ming; Mao Qiling, Lai Qijun, Yu Shida, and Wang Xu of the Qing; and Han Shaoxiang of the republic. In preserving Xiang Lake, they were saving the livelihood of those in the lake region. For centuries they were successful, until the world changed and made the lake seem conspicuously irrelevant.

But the final curtain has not fallen; the drama is not yet finished. The last act brings yet more violence; the denouement, the ultimate transmutation.

Postmortem: The Lake Region, 1955–1980

Several of the lines of Cai Panlong, the poet who depicted Xiang Lake in 1271, could summarize the political and social swirl of the quarter-century from 1955 to 1980: "Images, pictures come and go—one forgotten as it is replaced by the next. I am both amazed and confused." The reverses, the stops and starts of policy as the Communist party leadership felt its way along the road to a modern socialist state disoriented many of the people of Xiaoshan. After Mao Zedong called for intellectuals to speak out, to let a "hundred flowers bloom" in 1956, Jin Haiguan, the respected principal of twenty-five years of the Xiang Lake Normal School and longtime supporter of the revolution, did speak out about conditions in Xiaoshan. The next year he was condemned (along with 314 others in the county) as a "rightist," a label he carried to his death in 1971. He and the others were finally cleared of these charges in the aftermath of the Cultural Revolution in 1979.[41] Labels or names, as always, were defined by the regime in power, and as regimes shifted, the labels and names also changed, because a different present altered the definitions and "truths" of the past.

The years of the Great Leap Forward, 1958–1959, had significant effects on the county and determined many subsequent developments. Under Mao's leadership the Communist party moved the country from agricultural cooperatives to large-scale communes, military-style organizations in the countryside that, Mao contended, could marshal labor most effectively to increase production and move the country to "catch up" with Great Britain's level of production in fifteen years. In daily life the most visible symbols of communalization were the huge dining halls, constructed to free individual households from the time and labor of preparing meals and thus use their labor for the public more effectively. In Xiaoshan 2,726 dining halls, feeding 98 percent of the farm households, were constructed and operated from late 1958 to 1962.[42] The commune effort was another phase of the social revolution, which had begun with land reform, and a continuation of the twentieth-century trend of making private functions into public affairs.

Mao had also hoped to make every member of society feel a participant in both agriculture and industry—believing that a sense of participation would help increase production. This ideal plus an effort to decentralize industry led to the every-home-steel-smelter effort in which commune members took iron from home window frames and cooking pots and pans to contribute as scrap.[43] In the fervor of the steel production campaign, the Xiaoshan Steel Mill was established outside the county seat. From every unit and agricultural village, over 3,800 workers were drafted to work at the mill. But in a short time the mill simply closed: it lacked essential materials and sufficient technical expertise.[44] The Great Leap

Forward overleaped itself, toppling on its head. By autumn 1958 in another version of the "name" (or the "word") is the game, the county committee, like similar committees all over China, announced completely fictitious production statistics to hide the dimensions of failure.[45] For his part, Mao lost substantial personal power, and his commune policy began to be dismantled at the Lushan conference of the Central Committee in August 1959.[46] For the Xiang Lake area, however, it was clearly an awe-inspiring event when, following the conference, Mao came on an inspection visit to Xixing Commune, which contained land from the now-reclaimed Lower Xiang Lake. Playing the name game, the commune afterwards changed its name to The East Is Red Commune.[47]

During the economic debacle and social dislocations of the Great Leap Forward, Xiaoshan county underwent a surprising administrative change. Long before Xiang Lake was created, Xiaoshan had been an administrative unit of Shaoxing prefecture. Physiographically it was a part of the Ning-Shao Plain; in water drainage it was a part of the Three River system, made up of parts of Xiaoshan and Shaoxing counties. Historically, geographically, linguistically, it was a part of Shaoxing. On January 1, 1959, it became a part of the municipality of Hangzhou. The change permitted the provincial capital a much closer control of Xiaoshan. Since 1959 Xiaoshan has increasingly become the site of industry controlled and directed by the Hangzhou municipality rather than by the county or any of its units. Such factories, managed by Hangzhou men, are generally the largest in the county in total workers.[48] Further, in 1984 Xiaoshan was targeted specifically as Hangzhou's "industrial satellite municipality."[49] This arrangement means that Hangzhou can place industry far enough away to avoid some of the worst pollution problems and preserve its attractiveness for tourism. But as a consequence, in the mid-1980s Xiaoshan has become a grimy, dirty, polluted town. On many days through the gray-brown smog one can barely see North Trunk Mountain from West Mountain, little more than half a mile away; at the site of Chenxi Garden nature has been ambushed by a smoke-belching factory. While proximity to Hangzhou in the Song brought heightened importance to Xiaoshan as a contiguous county to the imperial capital, in the late twentieth century it has meant industrial proliferation with the consequent and increasingly grave burden of environmental pollution and decay.

The closer administrative relationship between Hangzhou and Xiaoshan also brought greater political violence to the county during the Cultural Revolution (1966–1976). Mao's attempt to restore the power he had lost after the Lushan conference, to perpetuate the ideals of class struggle and revolution among the young, and to insure the success of his "line" for development, the Cultural Revolution was Mao's campaign to rid the party of "capitalist-roaders" like Deng Xiaoping and Liu Shaoqi. It unleashed an orgy of violence and destruction that a decade later China was still decrying and from which, two decades later, it was trying to recover. In July 1966 the county established work teams to be sent to the rural communes to lead the Cultural Revolution. Red Guard units were established in schools, factories, and businesses. In the September campaign against the four "olds" (thought, culture, customs, and habits), 10,170 households in Xiaoshan were searched and their property confiscated: more than 558,300 yuan of goods were seized. Old books and paintings were burned or defaced; old buildings were destroyed. The attack on "capitalist-roaders" was taken to villages in former Xiang Lake in October. Less than forty years before, rural farmers had hidden from teachers of the Xiang Lake Normal School; though there is no account of the reception of the Red Guards, their advance into the countryside with a strongly political agenda points to the politicization of Chinese society in that relatively short period.

As the Cultural Revolution developed, Red Guard units began to form factions on issues of ideological purity and to break apart in increasingly emotional mutual accusations and outright bloody fighting. In the dead of night on January 20, 1967, two units of Hangzhou Red Guards crossed into Xiaoshan and attacked the Xiaoshan Public Security Bureau. Two weeks later, in early February, two opposing factions with links to Hangzhou were formed—the county General Union and the county General Command, the latter the more radical of the two. For the next seven months the county was immobilized by the continual struggle between these forces: factory production ceased; government organs were paralyzed. On August 26 the General Command and a radical faction of another Hangzhou unit launched a military attack on Xiaoshan, seizing the Public Security Office and its weapons and ammunition as well as the offices of the opposition. Fighting reportedly raged over the county: twenty-seven people were killed and countless numbers injured.[50] Less than two weeks later, the situation out of control, the People's Liberation Army moved in to establish military rule in Xiaoshan county.[51] Normal party rule was not reestablished in the county for twelve years (September 1979), and normal government rule, for almost thirteen years (June 1980). Although the date given for the conclusion of the Cultural Revolution in the nation at large is usually Mao's death and the arrest of the Gang of Four in September–October 1976, Xiaoshan continued to have open disputes between "revolutionaries" and "counter-revolutionaries" into late 1978.[52]

The 1980s: Re-creating the Lake

In many ways the 1980s are a throwback to the late 1940s. The destruction of the Cultural Revolution (like the earlier war with Japan) had passed. It has been a time to introduce new policies: in the 1940s people had searched for the reformist land policy that would bring effective change in the countryside. In 1980 Deng Xiaoping's vaunted new "responsibility system" allowed farmers to retain for themselves some of what they produced. Like the initial step of land reform in 1951, the "responsibility system" in the county was also tried out first at Long River, where the Lai lineage still dominated.[53]

The 1980s have also seen a return to "local rule." In the 1940s this phenomenon was evident in the provincial government's effort to reinvolve local leaders in regional affairs. In 1980 one sees for the first time since the Communist Revolution in 1949 county party and government leadership in the hands of people from Xiaoshan and adjacent counties. This change is particularly noteworthy because these leadership posts were almost completely monopolized from 1949 to 1980 by men from Shandong province. Since 1981 none of the thirteen key leaders have come from Shandong; only one has been from outside the province of Zhejiang (Jiangsu); and nine have come from counties that traditionally were part of Shaoxing prefecture.[54]

Even more extraordinary is that, like the late 1940s, the 1980s reopened the question of the future of Xiang Lake! It is altogether possible that the shift to a native, local leadership in the early 1980s played a role in this issue; that had happened previously, in 1926, when Shaoxing native Chen Yi became governor and temporarily gave the lake a reprieve. But the story of the transmutation of land to water, the rebirth of Xiang Lake, is more than the shift of official leadership. It finds its source in one of the dynamics in the history of the lake itself from the fifteenth century on —the brick and tile industry.[55]

At the end of the Qing dynasty, bricks and tiles were Xiaoshan's major product, each year's sales producing between 70,000 and 80,000 silver yuan. In 1925, before the effort at lake reclamation, sixty-three kilns in eleven lake villages made thirty-five kinds of bricks. In the late 1920s and the 1930s, many things undermined the once-flourishing industry: the reclamation of mud-dredging areas in Upper Xiang Lake, the resulting social disturbances between mud-dredgers and tenant farmers, the severe economic depression, and the consequent lack of demand for building materials. Japanese destruction and draconian control only exacerbated the decline. By 1949, no more than forty kilns were operating.

After liberation the area was in great need of building materials for reconstruction, and the government decided to support the restoration of the brick industry. By 1952 this industry was producing over 8 million bricks a year, double the total of 1949. Increased demand was spurred by a severely destructive typhoon in August 1956. From the mid-1950s on, the business expanded: more factories were organized by commune, county, and the Hangzhou municipality. In 1984 the brick factories produced almost 357 million bricks, eleven times what was produced in 1970, double that produced in 1976. Most of former Lower Xiang Lake and the northern section of Upper Xiang Lake have been excavated. The industry has flourished to a degree unpredictable in the economically troubled 1930s; in 1986 it was the province's foremost brick-making area.

The source of its prosperity, however, the clay that for centuries lay beneath Xiang Lake, is being depleted; production has begun to decline. Some quarries have already exhausted the clay resource: the Hangzhou Brick Factory pit is at fifteen meters and can go no further. Many quarries are already forty meters deep. The usable clay will be exhausted in all areas when the depth of fifty meters is reached. With the end of the industry in sight, in 1984 the county government set forth a plan to restore both Xiang Lake and its natural surroundings. Whenever the clay is exhausted (perhaps in little more than ten years) the quarries and pits will be filled with water from the Puyang and Qiantang rivers through the Little Li Mountain watergate. In most areas of Lower Xiang Lake and some areas of Upper Xiang Lake, it will reach a depth of close to fifty meters, more than the greatest original lake depth.[56] The new Xiang Lake, like the old, will be put to the practical use of feeding the area, this time through the scientific raising of fish rather than for irrigation.

For some, a decade or so is too long to wait to re-create the lake. On June 29, 1986, a severe flood washed away dikes along the forty-meter-wide canal in Lower Xiang Lake, destroying much of the Hangzhou Cogwheel Factory and filling one of the largest clay quarries near Chrysanthemum Mountain (see map 14). The water-filled quarry initiated a controversy redolent of past debates between those who wanted the quarry to remain lake and those (the brick industry) who wanted the quarry pumped dry. In this case, the brick industry won; by October the quarry was free of water. It is likely that the lake creators backed down because of the commitment in the foreseeable future to its reestablishment.[57]

As for restoring the surroundings of the future lake, efforts have already begun to build pavilions, teahouses, and viewing sites on certain mountains around the lake. West Mountain was developed as a park by the county government in 1981. Stone steps snake up the mountain and connect pavilions ornamented with traditional symbols and designs and painted in garish reds, blues, and greens. Though the buildings lack the beauty and the likely subtleties of design and decoration of the mountain's long since destroyed Shrine for the Virtuous and Kind or the Pavilion for Viewing the Lake, the park is designed not for the musings of retired scholar-poets but for families of picnickers and sightseers in the people's state.

At Xiang Lake there are three kinds of time. The first, the unbroken chain of linear time depicted in the Xiang Lake Normal School poem, points to the continuity of each event with the past, the structuring of each event by the past. A new wave forms in the context of the old wave's shape and intensity. Encroachments of the past, whether petty or widespread, became the context for present encroachments. Planting narrow strips of soil or building a kiln at water's edge transmuted lake to land and served as the base for the next transmutation. The lake's demise resulted from such cumulative encroachments linked in linear time. It was this kind of time that concerned lake preservers and reformers and that was the focus of lake historians: time—from beginning (1112) to end (circa 1955), from birth to death.

It is this kind of time that makes the rebirth of the lake remarkable. Yet two other kinds of Xiang Lake time suggest that it is instead to be expected. The first is the eternal cosmic flux of yin and yang that alternates over time in the "field of human events." As the time of yin waxes, that of yang wanes; but the time comes when yang begins to wax and yin to wane. The long-term decline of the lake system was set in motion by Sun Xuesi's construction of the Cross Lake Bridge to further the success of his brick-making enterprise. That began a long period of the waxing of the yang of kiln fires over the waning yin of lake water. Over the years, the health of the industry and of the lake as reservoir seemed to vary inversely with each other; as part of one system, each was intimately linked to the other. The expansion of the industry into lake territory in the seventeenth and eighteenth centuries seemed to hasten the lake's deterioration. In the early twentieth century the industry flourished as its dredging policies further damaged the lake. Finally, the period of Xiang Lake's death has been the industry's most prosperous. But time (and nature) have now brought the industry to the brink of its own destruction and, in yin and yang fashion, the industry is being replaced by the lake.

The third kind of time is cyclical—paradigmatic Xiang Lake time. Cycles of planting and harvesting endlessly repeated formed the very substance of Xiang Lake farm life. The wealth of lake poetry celebrated the cycle of seasons from Cai Panlong's panegyric on the lake to Wei Ji's cycle of poems on the seasons to Tang Boyin's 1949 praise of the lake's rusticity. The cycle of the transmutation of one traditional Chinese element, water, to another, earth, and back again will soon be complete. The story of the lake (and if we include the existence of the earlier natural Lake West of the Wall) is one of cycles of birth, death, rebirth, death, rebirth—a Buddhist parable of the nature of life itself. These latter kinds of time are the domain of the artist, the poet, and the religious (men like Lai Jizhi of the early Qing dynasty).

If the first kind of time at Xiang Lake is the dimension of the tragic—death and devastation by invasions, self-interested men destroying others, the rampages of nature—then the latter kinds of time are the dimensions of hope, endurance, and acceptance. Surely it must have been in part the strong sense of the cyclical order of things and the natural flux of yin and yang that enabled those around Xiang Lake to withstand centuries of tragedy.

Near the lake site in 1986 are two symbolic evidences of the lake's coming life cycle. The first is that the water-shield plant, long a lake symbol but eradicated in the late nineteenth century, has been recultivated. In 1980 at Old Tiger Cave Village near Wen Dike a special water-shield cultivation brigade had its first success in cultivating it for export. In 1984, 49.7 mu (about 8 acres) produced fifteen tons.[58] There is no word that the water-shield will necessarily be replanted in new Xiang Lake, but its revival in conjunction with the decision to re-create the lake is notable. The second symbol must be seen in the context of the wholesale destruction of the historical pavilions, shrines, and gravesites around the lake during the twentieth century. Although Wei Ji's grave at the foot of Xujiawu, east of the lake, has also been destroyed, there still remains before the old gravesite a five-meter-wide, thirty-meter-long grave path, on either side of which stand crumbling stone men, horses, sheep, and tigers.[59] In contrast, one thinks immediately of the similar gravesite structures of Sun Xuesi, demolished by his descendants out of shame for Sun's actions. Wei, who evinced great concern for the lake in actions and in poetry, is an obvious symbol of the lake: while his gravesite structures remain some five centuries later, the lake is being re-created. In contrast, the brick empire built by Sun is now being destroyed by his descendants and by nature itself.

In this present cycle, lake water, with its wealth of fish and water plants, its beauty reflective of the heavens, and its changing face in the four seasons, will again cover much of the area. In autumn 1986 water control engineer, Xiang Lake Normal School teacher, and county history editor alike mentioned their great satisfaction and pleasure over the lake's future re-creation. In flights of fancy, one can perhaps imagine that in a benign ghostly presence at Pleasant Hill, Wei Ji, with his life-filled vision of Xiang Lake, is smiling at the outcome. And, if we could see the wraiths of Cai Panlong and Sun Dehe watching the lake from the pavilion for seven often tragic centuries, can we doubt that "their ancient, glittering eyes" are gay?

Epilogue

One summer day at the end of the twentieth century, I climbed the King of Yue Garrison Mountain. It was not a big feat; countless others had done it over the centuries. Throughout history many climbers had put brush to paper or silk and expressed their reactions in poetry or in painting. As I climbed, I especially remembered Magistrate Wu Shu's rather detailed account of his ascent of the mountain in the 1480s (Chapter 3, "View"). There were still similarities with the mountain as it had existed then, but on that summer day the past was hard for me to see, for words from an earlier briefing in the Xiaoshan county government office echoed in my ears: "Come and invest." Given China's long history and its traditional relationship to outsiders, the words had an ironic ring. China's attitude to foreigners in the past had been best expressed by the words "Come and be transformed." The Chinese assumed that the China experience would transform the outsider through his or her contact with a superior civilization. Now county government officials were calling on outsiders to help transform Chinese civilization.

The county official's three words indeed showed how much China had changed not only since the 1480s but even since the 1980s! The reforms put in place by Deng Xiaoping in the early 1980s have surged forward into a full-fledged economic revolution, carrying China, and especially Xiaoshan and the Xiang Lake region, into uncharted territory. By whatever name they are known—capitalism or for the Chinese regime the more politically correct "socialism with Chinese characteristics"—they have already transformed Chinese society.

At the turn of the twenty-first century, Xiaoshan County is one of the richest in China. Modern expressways link it to the provincial capital of Hangzhou to the west, with its Pizza Hut, McDonalds, and Kentucky Fried Chicken franchises and its discos. Whereas in the mid- and late 1980s, the trip from Xiaoshan to Hangzhou had taken forty-five minutes to an hour or more, one can now travel to the capital's neon jungles and luxury department stores in fifteen to twenty minutes. Then it's on to Shanghai in just another two and a half hours by superhighway. Until 1937, Xiaoshan had been linked to Hangzhou only by ferry across the Qiantang River. The one bridge both completed and destroyed in 1937 was repaired after Liberation, but no other bridges spanned the river until the 1990s, when two more suspension bridges were constructed north of the original. Modern expressways also tie Xiaoshan to the ports of Ningbo and Wenzhou on the East China Sea and to the interior city of Jinhua. Xiaoshan city boasts two train stations, one for the Shanghai-Hangzhou-Ningbo line and one for the Zhejiang-Jiangxi line. In short, the 1990s saw Xiaoshan (and the Xiang Lake region) tied ever more closely to centers in Zhejiang province and beyond.

Even more significant for the image of the area was the construction of the new Hangzhou International Airport in Xiaoshan, only six or seven miles from Xiang Lake. The landing of jets from around the world bringing passengers who might very well "come and invest" is a symbol of the spectacular changes that have almost completely remade the area since the 1980s. Xiaoshan is thus being linked to the world beyond China. Little wonder that the turn-of-the-century president of the People's Republic and chair of the Communist Party, Jiang Zemin, has visited. In startling contrast to the visit of Mao Zedong to the East Is Red Commune in 1959 to herald the so-called realization of Communism, Jiang came to hail, if not in name, the victory of capitalism in transforming the area beyond recognition from just two decades before.

The development of the city of Shaoxing, the 1645 Ming loyalist capital in the holdout against the Manchu invaders, underscores the revolution bringing this region into the world economy. Between 1992 and 1996, its foreign trade increased an average of 30 percent a year. China Textile City between Xiaoshan and Shaoxing is the largest wholesale textile market in Southeast Asia. Yet the most astonishing statistics reflecting the economic boom occurred in 1997 when the flourishing construction industry created between 150,000 and 200,000 new jobs in Shaoxing (a city of four million). There can be little surprise that cities are flooded with masses of people from rural areas, the so-called floating population.

The most tangible evidence of the economic spectacle is in housing. Farm homes in the 1980s were still often at the level of hovels. Those near Xiang Lake were dark and cramped and had adobe or brick walls, and, those away from the lake, toward the coast, were similar structures with thatched roofs. By the late 1980s, farmers, reaping the fruits of changed government economic policy, had begun to construct two-story homes of plain brick. By the mid- and late 1990s, farm homes had become almost palatial: tiled three-story homes of intricate design featuring colorful roofing and decorative tiles, many windows, occasional balconies, spacious rooms, modern appliances, air conditioners, VCRs and DVDs, stereo systems, garages, yards, and satellite dishes. These housing developments rivaled any seen in U.S. suburbs, and they revealed the birth of a consumer culture driven by the desire for material acquisitions.

And what of Xiang Lake and of the plans made for it before the explosion of the economic revolution in the 1990s? The 1984 county government plan called for the former lake, which had been turned primarily into quarries, to be restored as a lake. But, as was seen in the aftermath of the severe flood of 1986, the brick and tile industry could still have its way with authorities—when it succeeded in having the large quarry near Chrysanthemum Mountain pumped dry of floodwater. One thing is certain: The days of brick- and tile-making at Xiang Lake are numbered as the clay resources continue to diminish rapidly.

According to new plans, Upper Xiang Lake will remain primarily land crisscrossed by canals. Of the 17,000 mu (2,720 acres) in Lower Xiang Lake, only 3,000 mu (480 acres) will become water space, and rice paddies will constitute about a quarter of the area (4,000 mu). The remaining 10,000 mu (1,600 acres—or almost 60 percent of the total) will be nonagricultural land. Some of this land will be used for the construction of terraced villas in a development called Xiang Lake Gardens: Members of the Xiaoshan economic commission traveled to Hong Kong in the mid-1990s to market this real estate opportunity. As an indication that the developers and marketers would target older retired Hong Kong residents desiring homes surrounded by natural beauty, some land was given over for the construction of health care facilities. Much of the land will go to the establishment of recreational opportunities, such as shooting clubs and fishing facilities, and to the enhancement of sightseeing possibilities for tourists. The establishment of the West Mountain Park in 1981 was the first such effort, made well before current plans were set forth. More important for Chinese aware of the area's history was the restoration of sites such as the King of Yue Garrison Mountain.

Let us now return to the climb of this mountain, named for the famous monarch who defended it in 473 B.C., defeating his chief enemy, the king of Wu. Unlike Magistrate Wu in the 1480s, I had no sedan chair to carry me into the thick forest on the mountain, but I also did not have to deal with his "dangerously steep" climb. Under the plan for tourism, the restoration of the mountain sites has provided a path with stairs to facilitate the climb to the top. At the beginning of the path is a stone arch inscribed, "Meditate on the Past of Garrison Mountain." In light of the rapid change in the area, this arch and its inscription are interesting for several reasons. First, it strikingly reminds us of the traditional practice of erecting such memorial arches, which often became objects of attack during Mao's Cultural Revolution. Second, in light of the Communist condemnation of the past as feudal, the call to meditate on the past underscores the very different kind of regime in power today. Finally, the character used for the word "meditate on" *(huai)* is written not in the simplified form according to the system established by the Communist regime but in the more complex form still used today in Taiwan. During the reform period it has become a fad of sorts to sprinkle nonsimplified characters through advertisements and other forms of public writing. In short, the arch at the very beginning of the climb opens not only the mountain but also a window on a China seemingly willing to give up at least the cultural trappings of communism and turn its eyes once again toward elements of traditional culture.

The county has constructed a pavilion at midpoint along the climb, a "little half-way house"—to use the image from William Butler Yeats's poem "Lapis Lazuli." From the pavilion on that brutally hot afternoon, like the three Chinese in Yeats's poem, the head of the county gazetteer committee, Mr. Chen; the history professor from Zhejiang University, Mr. Que; and I looked out "on the mountain and the sky." From there the vistas of Xiang Lake should have been magnificent: small bluish lakes and ponds separated by chunks of land of varying size; chartreuse rice paddies; the villages, clumps of white houses with red-tiled roofs studded with the smokestacks of kilns. The vistas *should* have been magnificent, but the sad truth was that the air pollution was so dense that we could barely see what stretched beneath us. As the reforms have progressed, so has the immense problem of environmental degradation continued to worsen.

We climbed on until we reached the top. At places the mountaintop was overgrown; if this was to appeal to tourists, it would be to those who like their nature wild. Vines and weeds grew over an old stone wall, somewhat reminiscent of Magistrate Wu's description of the stone cliff covered with old vines. Foliage mostly hid the sign identifying (in both Chinese and English) the Washing Horse Pond. In Magistrate Wu's day it allegedly produced freshwater fish and shrimp, but at the turn of the twenty-first century it was covered mostly with greenish scum, accented in places by lily pads, their white flowers revealing deep pink centers. Banks of greenery—trees, bushes, and vines growing in uncontrolled profusion—surrounded the pond.

Farther to the west on the top of the mountain was a newly built temple in honor of the king of Yue. A quarter of a century earlier, such a commemoration of a feudal king would have been unthinkable. Now a celebration of this man was a toast to the identity of the region known historically as Yue, today one of the wealthiest in China but significant for many reasons for over two millennia. Inside the temple was a large statue of a seated king of Yue, flanked by two attendants. Made of plaster and painted an ugly sallow yellow, they appeared quickly made and overly stylized rather than created with attention to individual details of face, body, and clothing—too commercial by half. They were in a sense a symbolic wedding of traditional history and values with the capitalist commercialism of the present, the very result that officials were now seeking to achieve in their plans for Xiang Lake.

Descending the mountain surrounded by swarms of mosquitoes, we came once again to Sun Village on the Lake, where we had begun the climb. Over five hundred years ago, when Magistrate Wu climbed the mountain, he visited the same village and was welcomed by a villager over ninety years old. When we descended, the number of children in the area—playing tag, throwing balls, running and squealing with delight—was remarkable. They were growing up in a new China, both filled with new promise and fraught with immense problems, but it was a China that seemed to be more open to the past than in prior years. Although the plans of the 1980s to remake the lake have changed, and though restoring sites around the lake may have a monetary objective, the past is no longer considered feudal.

Throughout its history, the story of Xiang Lake is a reflection of the development and values of the larger society in which it functioned as reservoir, as source of inspiration, as source of profit, and as an expression of its spatial context. Over the centuries, like China itself for the Chinese, the lake has been at the center of thinking for the lake community—whether farmers, fishermen, brick and tile makers, scholars, or poets. Cai Panlong said it as early as 1271: "Xiang Lake's unrestrained, heroic spirit pervades all; and all this is pressing in on my eyes here in this pavilion, fixing my gaze." The 1984 plan to remake the lake promised to restore this lake-centered reality.

But the reforms of the 1980s and 1990s seemingly have forever changed this possibility. Now land and homes in the lake region are marketed to people in Hong Kong and elsewhere. As superhighways tie Chinese cities ever closer, and as jumbo jets from every continent land less than ten miles away, bringing tourists and investors, the gaze of those involved with the lake has shifted beyond it to the wider world, just as the gaze of China has shifted in focus. If traditional lake-centered views made Xiang Lake, as the 1949 poem put it, "a plain village lass with cotton shirt," then the new beyond-the-lake views make it a high-powered, well-dressed, and well-coiffed businesswoman.

In the last chapter I spoke of three kinds of time associated with Xiang Lake: linear or chronological time, the flux of yin-yang time, and cyclical time. It was the latter two kinds of time that made the re-creation of the lake "explicable." In one sense, the turn of events brought by the economic revolution and the reorientation of the lake region obviates the meaning of those two kinds of time in the lake's history. That is not entirely the case, of course, because some of the original lake land will again become lake.

But in a very real way the change that occurred in the 1990s was a more radical alteration of lake-region reality than any that had occurred since the lake's creation in the twelfth century. More than a change in time, it has become a spatial or orientational transformation. Xiang Lake and its environs have been opened to the world beyond in ways never before imagined. How contact with that world will continue to transform the lake and its various communities will be the story of the twenty-first century.

Abbreviations Used in Notes

Diaocha	*Xiaoshan diaocha jihua baogao shu*
XC	*Xihe xiansheng zhuan* in *Xihe heji*
XK	Yu Shida, *Xiaoshan Xianghu kaolue*
XSZ	Mao Qiling, *Xianghu Shuili zhi*
*XX (*1693)	Xiaoshan xianzhi, 1693
XX (1985)	Xiaoshan xianzhi, 1985
XXG	*Xiaoshan xianzhi gao*
XXJJ	Han Yanmen, "Xianghu xiangcun jianshe jihua dagang cao'an," in *Zhejiang nongye tuiguang*
XXX	Zhou Yizao, ed., *Xiaoshan Xianghu xuzhi*
XXZ	Zhou Yizao, ed., *Xiaoshan Xianghu zhi*
ZJ	Xu Fanglie, *Zhedong jilue*

Notes

Preface

1. Such interpretations are commonplace. See, e.g., Kenneson, "China Stinks," esp. p. 18.

2. Naitō Konan set forth the division between the Tang and Song periods, denoting the Song as the beginning of the "modern" period of Chinese history. See Fogel, *Politics and Sinology*, pp. 179–82.

3. The nine views at Xiang Lake are a function not of authorial pretense but of the divisions of this particular history.

4. *XXZ*, 5:12a–14b.

5. *XXG*, 32:17b.

6. *XX* (1985), *Da Shiji*, p. 61.

I. VIEW: *From North Trunk Mountain, Late Twelfth Century*

1. *XX* (1693), *juan* 1: *pa jing*, 5:15a–b. Traditional bibliographies list the publication in 1683. It contains references, however, to a lake crisis in the late 1680s. The correct date is 1693.

2. *XXG*, 32:2b–3a. See also *Xiaoshan Zhao jiapu*, 8:1a–2a.

3. Kao, "Fang La Rebellion," p. 37. The rebellion's bloodiness was heightened by government soldiers who were rewarded according to body count tabulated in heads. Whole families were killed in their farmhouses by soldiers hungry for rewards.

4. Manicheanism had been introduced into China in the seventh century. See Kao, "Fang La Rebellion," pp. 55–61.

5. *XX* (1693), 5:3b.

6. Eberhard, *Local Cultures of South and East China*, pp. 349–62. Eberhard doubts that the figure Yu was primarily linked in the beginning to the Shaoxing area.

7. Gernet, *Daily Life in China*, pp. 195–96. For descriptions of the bore in the

nineteenth century, see Moule, *Half a Century in China,* pp. 128–33, and Moore, "Bore of the Tsien-tang Kiang (Hang-chau Bay)," pp. 188*ff.*

8. *XXG,* 7:3b–4a.

9. Shiba, *Commerce and Society in Sung China*, p. 9.

10. Sir Henry Yule, trans., *The Book of Ser Marco Polo*, ed. Henri Cordier, 2:185.

11. Chen Qiaoyi, "Lun lishi shiqi Ning-Shao pingyuandi hupo yanbian," p. 29.

12. *Xiaoshan Zhao jiapu*, 8:1a–2a.

13. Mihelic, "Polders and the Politics of Land Reclamation," p. 210.

14. *XXG,* 3:49b–55a; see also *XXZ,* 5:12a.

15. It follows that a building should face south for a favorably auspicious geomantic location. For a description of the traditions of the Dual Forces (yin-yang) and fengshui, see Wright, "Cosmology of the Chinese City," pp. 41–56.

16. *XXZ,* 5:4a–5a.

I. CHAPTER: *The Beginnings, 1112–1214: Four Servants of the People*

1. *XXZ* 5:11a–12a. Professor Mark Schwehn of Valparaiso University first pointed out the dramatic parallel between this scene and the one described by William Butler Yeats in "Lapis Lazuli." My account of Xiang Lake resonates with themes and images from this poem.

2. *XSZ* 1:1a–b. See also *XXG*, 3:28a. A county was the smallest administrative unit in a province, the lowest-reaching extension of the state bureaucracy. Counties were divided into townships of varying sizes; in Xiaoshan during the Song, they probably averaged roughly four to seven square miles.

3. This account is based on a fascinating essay by Chen Qiaoyi, "Lun lishi shiqi Puyangjiang xiayude hedao bianqian," pp. 65–79.

4. *XXG*, 3:30b.

5. McKnight, *Village and Bureaucracy,* p. 182. The evolving civil service system had not yet become firmly established as a route to such status, even though the power of the old aristocracy had been broken.

6. Stuermer, "Polder Construction and the Pattern of Land Ownership," p. 202.

7. McKnight, *Village and Bureaucracy*, p. 24. The account of the 1112 meeting is in XSZ, 1:1b–2b.

8. *XXG*, 3:51b, 52b. Following Maurice Freedman, I define *lineage* as a "permanent organized group" tracing "patrilineal descent from one ancestor." See Freedman, "Introduction," in *Family and Kinship in Chinese Society*, pp. 13–14.

9. The sources indicate that the circumference of the original lake was 82.5 li; however, the li during the Song and Ming was only .83 of a li during the republican period, when 3 li = 1 English mile. Therefore, the original circumference in republican li was 68.5 li, or 22.8 English miles. See *XX (*1985), *Fulu*, p. 13.

10. The tax was called the *junbao humi*. It amounted to 7 *ge* and 5 *shao* per mu (a *ge* is 1/10 of a pint; a *shao* is 1/10 of a *ge*). The cropland turned lake land produced 1,000 piculs, 7 pints, and 5 ge. We are not told what the base year was (*XSZ*, 1:1b–2b).

An acre is 6.6 mu of land. Mark Schwehn's suggestions were significant in handling this section.

11. *XK*, 1a–2a. For the amounts of water to be distributed to the number of mu in the townships, see *XSZ*, 1:6a–10b.

12. Shiba, "Sekkō Shōsanken Shōko," p. 291. For a description of Song dynasty household grading, see McKnight, *Village and Bureaucracy*, pp. 127–30.

13. *XXG*, 12:1b–2a.

14. *XSZ*, 1:3a.

15. *XSZ*, 1:6a–10b; *XXZ*, 5:6b.

16. *XSZ*, 1:5a–b.

17. *XSZ*, 1:3a.

18. Once before, in 1119, one or more *local* strongmen (*hao*) had attempted to reclaim part of the lake. See *XSZ*, 1:2b–3a. Although many opposed this 1119 reclamation, the proponents were from more powerful and prestigious "resident households." (For "resident" and "guest" householders, see McKnight, *Village and Bureaucracy*, p. 126.) It may be that these reclamation proponents also had ties to higher-level officials, because their efforts coincided with a reclamation program north of Hangzhou Bay. Such ties would have strengthened their position. See Mihelich, "Polders and the Politics of Land Reclamation," pp. 184–88, for information on the program.

The issue was resolved by one Elder Bian, a local leader with great prestige. Drought and the necessity of irrigation, he ruled, made reclamation unwise and unwarranted. Records reveal no reaction by reclamation advocates to the decision, quoting instead a popular saying of the time: "The people have heaven; the lake is not used for paddy land" (*min you tian, hu bu tian*).

19. *XX*(1693), 15:4a–5a.

20. *XSZ*, 1:3a–b.

21. *XSZ*, 1:3a–b. There is some debate over the details of the Wang Qu encroachment. One source records that the unnamed man near the lake took the initiative by providing the lake turned paddy as a gift to Wang. His motive is only speculative. Another source places the onus on Wang for his greed in arranging the whole affair. See *XXG*, 3:28b.

22. *XSZ*, 1:4b–5a.

23. *XXG*, 3:50b.

24. *XX* (1693), 7:1a; Chen Qiaoyi, "Lun lishi shiqi Puyangjiang xiayude hedao bianqian," pp. 33–34.

25. Stuermer, "Polder Construction and the Pattern of Land Ownership," pp. 75, 83, 87, 89, and 95, n. 13. See also discussion of the population increase in Hymes, *Statesmen and Gentlemen*, p. 1.

26. Chen Qiaoyi, "Lun lishi shiqi Ning-Shao pingyuandi hupo yanbian," p. 36. See the similar analysis in Will, "Occurrences of, and Responses to, Catastrophes and Economic Change," p. 6. See also Mihelic, "Polders and the Politics of Land Reclamation," esp. chap. 5, for the reclamation of Lake Guangde.

27. *XSZ*, 1:3b–4b.

28. This account is based on *XSZ*, 1:10b–11b.

29. The two townships were Yuhua and Xiaxiao.

30. *XXG*, 3:49b–50a.

31. The petition noted that perhaps out of a guilty conscience or to cover his wrongdoing, Zhang gave large sums of money to relieve the plight of famine sufferers.

32. Shaming as a part of public policy was used in the Cultural Revolution in the parading of political opponents in dunce caps.

33. Included on the placard, and suggesting that the social situation around White Horse Lake had become quite volatile, were cautions against social disturbances and wild talk coupled with calls for public spiritedness.

34. *XXG*, 3:50a–51b, 52b.

35. *XXG*, 3:50a–51b, 52b. Zhang's "end run" was probably prompted in part by knowledge that Zhu Zhaoling, a judge of the provincial court, had been apprised of the damage done to the lake.

36. *XXG*, 3:51b–52b. The official to whom the lake was presented was Zhao Chengxuan.

37. *XXG*, 7:2a, 3b–4a.

38. *XXG*, 3:52a–b.

39. *XXG*, 3:52b.

40. *XXG*, 3:51a.

41. *XSZ*, 1:3b–4b.

42. *XSZ*, 1:3b–4b. For Hymes's account of the roles of the state and the local elite in waterworks, see *Statesmen and Gentlemen*, pp. 167–76.

43. The following account is based on *XSZ*, 1:11b–12a.

44. *XXZ*, 8:14b.

45. See Mihelic, "Polders and the Politics of Land Reclamation," p. 211.

46. *XXZ*, 8:14a–b.

47. *XXZ*, 1:2a.

48. See the poems by Lou Like and Liu Huan, *XXZ*, 7:1b, 2a.

49. See Shi Rulan's poem, *XXZ*, 7:1a–b.

50. *XSZ*, 1:12a.

51. Though the heyday of Buddhist success in China is generally said to have ended in the cataclysmic persecutions of the 840s, in Zhejiang province, judged by the number of temples reconstructed or newly built, a Buddhist revival began as early as the 860s. The position of Buddhist institutions was strengthened during the Five Dynasties period (907–960) by the beneficent patronage of the rulers of the state of Wu-Yue. See, e.g., *XXG*, 8:1a–28b, and XX (1693), juan 2. This patronage continued under the Song, and the Buddhist establishment, especially in Hangzhou, became large landowners in the area. See Mihelich, "Polders and the Politics of Land Reclamation," pp. 218–19, and Hymes, *Statesmen and Gentlemen*, pp. 8, 178–79. It frequently vied with other elites from the capital to recover lake land for private use. In 1182 a monk from a nearby temple asked for and received over 120 acres of lake land from Dinghai county to the east (*XXG*, 3:51a). A monk from Hangzhou asked for and received land from Shooting Star Lake in 1198; because of a series of petitions from township households near the lake, the paddy was restored to lake in 1200 (*XXG*, 3:51b). In the period 1228–1234, a local monk turned a portion of Xiaoshan's Tong Lake to paddy (*XXG*, 3:53b).

52. *XXZ*, 7:5a.

II. VIEW: *Chenxi Garden, Late Fifteenth Century*

1. See the description in Ayscough, "Chinese Idea of a Garden," pp. 15–22.

2. See Ayscough, "Chinese Idea of a Garden," p. 18; see also Morris, *Gardens of China*, p. 73 For a description of the destroyed city walls, see *Shaoxing fuzhi*, 2:5b.

3. *XXG*, 33:3a–b.

4. Morris, *Gardens of China*, p. 165.

5. This description follows Morris, *Gardens of China*, pp. 170–86, and *XXG*, 1:35b–36a.

6. *XXZ*, 5:12a–14b.

7. *XX* (1751), 32:1a–8b; *XXG* (1935), 8:7b–8b.

8. Morris, *Gardens of China*, pp. 185–86.

9. The epithet "kingfisher blue" was a stock phrase in the poetry detailing the lake's scenery. See the poetry in *XXZ*, juan 5–7.

10. Morris, *Gardens of China*, p. 186.

11. *XXG*, 8:14a.

12. Morris, *Gardens of China*, p. 180.

13. *XXZ*, 8:10a.

II. CHAPTER: *Obligation and Death, 1378–1500*

1. *XXG*, 15:3a.

2. *XXG*, 15:3a–b.

3. See Dennerline, *Chia-ting Loyalists*.

4. Dennerline, *Chia-ting Loyalists*, p. 263.

5. *XX* (1693), 3:3b–4a.

6. See Fang's biography in Goodrich, ed., *Dictionary of Ming Biography*, 1:433–35.

7. *XXG*, 32:2b–3a; *Xiaoshan Zhao jiapu*, 8:1a–2a.

8. The account is from Xu Mianzhi's diary account, *Bao Yue lu*. The section dealing with Xiaoshan is mostly printed in *XXG*, 11:5a–b. See also the account of siege warfare in Shaoxing in Franke, "Siege and Defense of Towns in Medieval China," pp. 188–92.

9. *XXG*, juan 5.

10. This account is from *XSZ*, 1:12b.

11. *XXG*, 12:4b, is a brief biography of the judicial officer Cui Jia.

12. *XXG*, 12:4b–5a.

13. *XSZ*, 1:13a.

14. See *XXG*, 1:27b; *Shaoxing fuzhi*, 12:12b; for the Double Nine Festival in Hangzhou during the Song, see Gernet, *Daily Life in China*, p. 196; Morris, *Gardens of China*, p. 186.

15. For the biography of Wei, see *XXG*, 14:7b–8a. For the early Ming practice of assigning failed jinshi degree seekers to such posts, see Ho, *Ladder of Success in Imperial China*, pp. 26–27.

16. Goodrich, ed., *Dictionary of Ming Biography*, 2:1460.

17. For this poetry, see *XXZ*, juan 7.
18. *XXZ*, 8:8b.
19. *XXZ*, 7:3b.
20. *XXZ*, 7:4a.
21. *XXZ*, 7:4a–b.
22. *XXZ*, 7:3a–b, 4b.
23. *XXZ*, 7:2b–3a.
24. *XXZ*, 7:4b.
25. *XSZ*, 1:16a–b.
26. *XSZ*, 1:14b.
27. *XSZ*, 1:14b–15a.
28. *XSZ*, 1:14b–15a.
29. *XSZ*, 1:15a–b.
30. *Xiaoshan xiangtu zhi*, p. 8. See also *XXG*, 3:22b–23a, and *Zhejiang shuiliju niankan*, pp. 7–8. Chen Qiaoyi contends that Qiyan was excavated as early as the late twelfth century and that the opening was alternately diked and reopened until its final reopening in the Tianshun reign. See "Lun lishi shiqi Puyangjiang xiayude hedao bianqian," p. 75.
31. *XXZ*, 3:7a–8b.
32. This account is from *XXG*, 3:32a–b.
33. *XSZ*, 1:15b.
34. The following account is based on *XSZ*, 2:1a–3a, 14a–b, and *XXG*, 14:8a–9a.
35. *Shaoxing fuzhi*, 9:22a–b.
36. For a biography of Tong, see *XXG*, 14:9b–10a.
37. Tong served as educational officer in various posts throughout his career. He died at age seventy-nine.
38. The following account is based on *XSZ*, 2:1b–2b, and *XXG*, 14:9a–b.
39. The analect is from Book 13, number 18: "The 'Duke' of She addressed Master K'ung saying, 'In my country there was a man called Upright Kung. His father appropriated a sheep, and Kung bore witness against him.' Master K'ung said, 'In my country the upright men are of quite another sort. A father will screen his son, and a son, his father'" (Waley, trans., *Analects of Confucius*, pp. 175–76).
40. *XXG*, 9:14b.
41. The phrase is from a poem by Fu Yiyin, *XXZ*, 7:36a.
42. *XXZ*, 7:36b.

III. VIEW: *The King of Yue Garrison, Late Sixteenth Century*

1. This account is based on Maspero, *China in Antiquity*, p. 219; see also Hirth, *Ancient History of China*, pp. 233, 261, 348.
2. From the *Shiji*; quoted in Maspero, *China in Antiquity*, p. 220.
3. Quoted in Moore, "Bore of the Tsien-tang Kiang," p. 232.
4. Moore, "Bore of the Tsien-tang Kiang," p. 232.
5. *XXZ*, 5:12a–14b. Note also the short poem by Zhu Yuzhen, *XXZ*, 7:11b, for a description of climbers of the mountain a millennium after Kouqian. For a shorter

poetic description of a trip up the mountain, see Lai Cengyi, *XXZ*, 7:12b. See also the poem by Zhang Yuan of the Qing dynasty, 7:26b.

6. See, e.g., the poems by Ming dynasty scholars Lai Sanping, *XXZ*, 7:8b, and Ding Kezhen, *XXZ*, 7:13a, and by Qing dynasty figures Chen Zhiyin, *XXZ*, 7:22b, and Han Suizhi, *XXZ*, 7:28b–29a.

7. The story is recounted in *Xiaoshan xiangtu zhi*, pp. 55–56.

8. *XX* (1693), 15:1a.

9. See *XX* (1693), 15:4a–7a.

10. For the negative views on Buddhist clergy, see Ch'en, *Buddhism in China*, chaps. 14–16.

11. *XXZ*, 7:16a.

12. *XXG*, 32:18a.

13. *XXZ*, 7:14a.

III. CHAPTER: *Strange Fruit: Lineages and the Lake, 1519–1555*

1. On portents, see Burton Watson, *Ssu-ma Ch'ien*, pp. 98–100.

2. *XXG*, 5:25a–b.

3. Sources do not agree on the date of construction. Compare *Shaoxing xianzhi ziliao diyi ji*, 5:15b, which gives 1536; *Zhejiang shuili ju niankan*, which gives 1537; and *Xiaoshan xiangtu zhi*, which gives 1539. The three rivers were the Qiantang, Puyang, and Cao'e.

4. *XXG*, 3:20b.

5. *XXZ*, 2:18b–20a. The number of mu was found by adding the amount from the Song for the first three; taking 20 percent of the Song amount for Anyang and Xuxian; and taking the number of mu for Laohu Village in Yuhua Township, still getting water from the lakes, at least sporadically.

6. These arguments are based on actual arguments by reclaimers and reports of lake violators from the late Ming to the twentieth century. See *XXZ*, esp. juan 1–4.

7. See the analysis of Mancur Olson in *Logic of Collective Action*, chap. 1. Olson argues that "the larger the group, the less it will further its common interests."

8. *Xiaoshan xiangtu zhi*, pp. 43, 45.

9. See the discussion in Hayes, "Specialists and Written Materials in the Village World," pp. 80–81.

10. On ancestral halls, see Hazelton, "Patrilines and Development of Localized Lineages," pp. 151–54; on the role of genealogies, see pp. 164–65. On lineage estates, see James L. Watson, "Anthropological Overview," pp. 277–79, and Ebrey, "Development of Descent Group Organization," pp. 53–56.

11. It is likely that what I here describe as lineage interests are more accurately denoted as the interests of the leaders of one or a small number of patrilines from the larger lineage; cf. Hazelton, "Patrilines and Development of Localized Lineages," p. 166.

12. The account of Sun's and Wu's activities is in *XSZ*, 2:5a–7a. For the rebellion see Goodrich, ed., *Dictionary of Ming Biography*, vol. 2: biographies of Wang Shou-jen and Liang Ts'ai.

13. See Morita, *Shindai suirishi kenkyū*, pp. 417–49. See also Elvin, "Market Towns and Water Ways," pp. 449–57.

14. On the establishment of lijia, see Watt, "Yamen and Urban Administration," p. 384. Civil service degree holders could not serve as lijia headmen.

15. See *XSZ*, 2:5a–7a.

16. *XXG*, 3:33b, has "within the first month of every year" rather than "every month."

17. See esp. the memorial of Gu Dingchen (1527), cited in Wiens, "Changes in Fiscal and Rural Control Systems," p. 64. See also Huang, "Fiscal Administration during the Ming Dynasty," pp. 110–12.

18. Elvin points out that in Shanghai the system of dike administrators was breaking down by the later sixteenth century, pp. 453–54.

19. *XXG*, 33:10a.

20. *Xiaoshan xiangtu zhi*, p. 14.

21. *Xiaoshan xiangtu zhi*, p. 43.

22. See, e.g., the Qing dynasty writings of Lai Jizhi, *XXG*, 32:18a, 20a; of Lai Congmin, *XXG*, 33:9b; of Lai Xiangyan, *XXG*, 33:9b–10a; and of Lai Congdao, XXG, 32:30b–31a.

23. *XXG*, 32:17b.

24. *XXG*, 15:6b.

25. *XXG*, 15:6a.

26. *XXG*, 15:5b.

27. *XXZ*, 7:5b.

28. See *Xiaoshan Daishang Huang shi jiapu*, 4:33a–b, 4:2b–4a.

29. *Xiaoshan Daishang Huang shi jiapu*, 4:5a–b.

30. On the significance of marriage strategies, see Dennerline, "Marriage, Adoption, and Charity," pp. 170–71, 181–86, and Hymes, "Marriage, Descent Groups, and Localist Strategy," pp. 102–13. See also Dennerline, *Chia-ting Loyalists*, pp. 133*ff*., for a discussion of the Hou family's marriage strategy.

31. See *XXG*, juan 13, 14. See Robert Hartwell's analysis on the role of the civil service examination in mobility and among local elites in his stimulating essay, "Demographic, Political, and Social Transformation of China," pp. 417–19.

32. The seven included two jinshi, two juren, and three gongsheng degree holders; the thirty-two were nine jinshi, eleven juren, and twelve gongsheng.

33. The fifteenth-century Sun degrees were one jinshi, two juren, and one gongsheng; the sixteenth-century degrees, one jinshi, three juren, and one gongsheng. The Wus' degrees were both gongsheng.

34. See *Xiaoshan Daishang Huang jiapu*, 4:23b.

35. For a general account of the situation, see So, *Japanese Piracy in Ming China*, and Fitzpatrick, "Local Administration in Northern Chekiang."

36. See Higgins, "Pirates in Gowns and Caps," pp. 30–36.

37. Fitzpatrick, "Local Administration in Northern Chekiang," pp. 72–73.

38. See *XXG*, 11:5b. The account of the building of the city wall by Magistrate Shi gives the date of this attack as 1552; other sources make it more likely the date was 1553. See XXG, 32:43b–45b, for three essays on the building of the wall.

39. *XX* (1693), 15:6b.

40. See *Xiaoshan Lai shi jiapu*, 1:27a–29a. Also see Fitzpatrick's descriptions in "Local Administration in Northern Chekiang," pp. 153*ff*. On the description of elite homes, see Moule, *New China and Old*, pp. 118, 123.

41. *XXG*, 15:11a. This generation of this particular branch of the lineage was prodigiously gifted and public-spirited: In addition to jiansheng degree holders Duanmeng and Duancao, who between them fathered five jiansheng degree holders, six others held jiansheng degrees and one a gongsheng degree. All were active following the pirate scourge in rebuilding bridges and water control facilities.

42. The accounts of these confrontations are based on *XXG*, 11:5b–6b (*bing shi*, "military affairs").

43. See, e.g., *XXG*, 2:23a*ff*., for lists of bridges needing repairing.

44. This account comes from *XSZ*, 2:14b.

45. For a general overview of this economic upsurge, see Rawski, "Economic and Social Foundations of Late Imperial Culture," pp. 3–11.

46. See Ho, *Ladder of Success in Imperial China*, p. 216, for discussion of recommendation as an alternative to rising through the examination system.

47. *XXZ*, 8:10b.

48. *XXZ*, 8:6b–7a.

49. *XXZ*, 8:10b.

50. *XSZ*, 2:14b.

51. See, e.g., his report on the dike in *Xiaoshan Daishang Huang shi jiapu*, 3:25a–30b; see also his biography in *XXG*, 15:9b.

52. *XXZ*, 7:7b.

53. *XXZ*, 6:1a.

54. *XXZ*, 1:11b–12a.

55. *XXG*, 32:24a–25a.

56. *XXG*, 32:24a–25a.

57. *XX* (1693), 11:8a*ff*.

58. Mao, *He yushi*, 10a–b.

59. *XXG*, 15:9b.

60. *XXZ*, 7:6a.

61. *XXZ*, 8:8b–9a. Grave images and decorations usually glorified the dead by representing significant events in their lives. See De Groot, *Religious System of China*, 2:826.

IV. VIEW: *On Pure Land Mountain, 1689*

1. *XXG*, 9:8b.

2. For an account of the temple, see *XXG*, 8:8b.

3. *XXG*, 33:8a.

4. *XXZ*, 8:10a.

5. See the discussion on such shrines in Yang, *Religion in Chinese Society*, pp. 161–77. See also Grimm, "Ming Educational Intendants," p. 143.

6. This description is based on *XXG*, 7:7a–8b, and *He Yushi*.

7. See Meskill, "Academies and Politics in the Ming Dynasty," p. 152.

8. Mao Qiling, *He yushi*, 3a–b.

9. See Meskill, "Academies and Politics in the Ming Dynasty," pp. 171–74, and Dennerline, *Chia-ting Loyalists*, pp. 217*ff*.

10. *XXG*, 32:1a.

IV. CHAPTER: *Image and Remembrance, 1644–1705*

1. *XXG*, 5:26a.
2. See, e.g., poems by Cai Zhongguang, *XXZ*, 7:17a–b.
3. *XXZ*, 7:18a.
4. *XXZ*, 7:17a.
5. *XXZ*, 7:19b.
6. *XXZ*, 7:9b–10a.
7. *XX* (1693), 20:19a–b. For an interesting account of suicide in Chinese culture, see Hsieh and Spence, "Suicide and the Family in Pre-Modern Chinese Society," pp. 29–47.
8. *XXG*, 14:13b.
9. *XXG*, 14:14a.
10. *XXG*, 15:6b. For the biography of Lai, see *XXG*, 15:20a. The two military officers reconciled by Lai were Ma Shiying and Zuo Liangyu.
11. *XXG*, 16:6a–b.
12. For Mao's biography, see *XC*. See also *XXG*, 16:8a–9a, and the biography in Hummel, ed., *Eminent Chinese of the Ch'ing Period*, 1:563–65.
13. *XC*, 11a.
14. The idea of a Daoist retreat into nature was deeply embedded in Chinese culture; during a long career in the frequently stultifying bureaucracy a scholar-official might take a sabbatical retreat far into the wilds of nature to write poetry, paint, or meditate.
15. *XXG*, 16:5a.
16. *ZJ*, p. 1. See also Hummel, *Eminent Chinese of the Ch'ing Period*, 1:492. The most extensive and thorough treatment of the Manchu conquest is Wakeman, *The Great Enterprise*.
17. *ZJ*, pp. 6–7.
18. *ZJ*, pp. 3–9 *passim*. For an account of the resistance in the south, see Wakeman, *The Great Enterprise*, 1: chap. 8, 2: chap. 10. See also Struve, *The Southern Ming*, pp. 77–96.
19. *ZJ*, p. 3.
20. *XXG*, 16:5a–6a.
21. *XXG*, 11:6b.
22. *ZJ*, p. 8.
23. *ZJ*, pp. 12–13.
24. *ZJ*, p. 11.
25. *ZJ*, pp. 13–14.
26. *ZJ*, pp. 13–14; see also Hummel, *Eminent Chinese of the Ch'ing Period*, 1:181.
27. *ZJ*, p. 16.
28. See below, p. 105.

29. This account is based on *XC*, 12b–13a.
30. *XC*, 13b.
31. *ZJ*, pp. 21–24.
32. *XXG*, 14:13b.
33. *XXG*, 14:14b.
34. *XXG*, 16:5b–6b.
35. Dennerline, *Chia-ting Loyalists*, pp. 286–87. Wakeman calls the triumph of the Manchus "the most dramatic dynastic succession in all of Chinese history" (*The Great Enterprise*, p. 1).
36. Hummel, *Eminent Chinese of the Ch'ing Period*, 1:563.
37. Yanwen died at the age of fifty, likely in the first decade of the eighteenth century (see *XXG*, 16:11b).
38. *XXZ*, 7:10a–11a.
39. *XXZ*, 7:10a–b.
40. *XXZ*, 8:8a–b.
41. *XXZ*, 7:10b–11a, 8:7b.
42. *XXZ*, 7:11a–b.
43. *XXG*, 15:20a, 16:11b–12a.
44. Zelin, *The Magistrate's Tael*, pp. 8–9, 313 n. 24.
45. *XXG*, 33:13b–14a.
46. This section is based on *XXZ*, 7:15a–18a.
47. *XXZ*, 7:15b–16a.
48. *XXZ*, 7:15a.
49. *XXZ*, 7:15a, 16a, 16b, 17a.
50. *XXZ*, 7:17b.
51. *XXZ*, 7:15a, 15b–16a, 16b, 17a.
52. *XXZ*, 7:17a.
53. *XXZ*, 7:15b.
54. *XC*, 13b–14b.
55. *Daoguang Guiji xianzhi*, 19:3a.
56. *XC*, 14b.
57. Hummel, *Eminent Chinese of the Ch'ing Period*, 1:563.
58. This account is based on *XSZ*, 2:7a–14a.
59. *XXZ*, 8:13a.
60. This account is based on Mao Qiling, *He Yushi*, 1a–23b.
61. There may have been some simple cupidity involved here, because the Wei lineage, keeper of the shrine, was able to keep money left over from the ceremonies. Additional monies for sacrifices to the Hes would decrease its income.
62. *XXG*, 7:11a.

V. VIEW: *Sailing on the Lake from Upper Sun Village to Green Mountain, 1783*

1. *XXZ*, 7:25b. For Cai's biography, see *XXG*, 16:14b.
2. *XXZ*, 8:14b–15a.
3. *XXZ*, 8:15a.

4. For the latter, see *Chinese Economic Bulletin* 21:10 (Sept. 3, 1932): 130. For mention of the buildings of molds, see *Xianghu diaocha jihua baogao shu*, p. 29.

5. *XXZ*, 8:24a–40b.

6. De Groot, *Religious System of China*, pp. 939–55.

7. *XXZ*, 8:10a.

8. De Groot, *Religious System of China*, pp. 941–42.

9. *XXZ*, 8:8a–10a.

10. *XXG*, 32:48a–49a.

11. *XXG*, 1:27a. The Festival of the Dead and the Cold Food Festival "express[ed] and strengthen[ed] agnatic kinship ties" (Hazelton, "Patrilines and the Development of Localized Lineages," pp. 152–53); see also Ebrey, "Development of Descent Group Organization," pp. 20–29.

12. *XXZ*, 8:10a.

13. *Xiaoshan Shi shi zongpu*, 1:n.p.

14. *XXG*, 5:27b.

15. *XXZ*, 7:26a. For his biography, see *XXG*, 16:14a–b.

16. *XXZ*, 8:9b.

17. Zhang describes himself in his poem as the filial son.

18. The description is from Moule, *New China and Old*, p. 110.

19. *XXZ*, 8:38b.

20. Moule, *New China and Old*, p. 124.

V. CHAPTER: *Restoring the Lake: Of Mud-Dredgers, Leeches, and Worm Officers, 1758–1809*

1. Albert Feuerwerker, *State and Society in Eighteenth-Century China*, p. vii.

2. This account is based on XXZ, 8:16b–24a. Only three of the eighteen were upper degree holders.

3. This generalization is based on the names of lineages in villages along the lake's east side found in XXZ, 8:24a–40a.

4. *XXZ*, 8:16b.

5. Worster, *Nature's Economy*, p. 2.

6. *XXZ*, 8:24a.

7. This account is based on *XXG*, 33:29b–33a.

8. *XXZ*, 1:28b–29a; see also *XXG*, 3:41a–42a.

9. See chap. 2.

10. This material is based on the recorded survey, *XXZ*, 8:24a–40a.

11. *XXG*, 3:41a–42a. Huang shows that such a survey had been completed in Zhejiang province in 1386. The "fish scale" nomenclature came from the "topographical charts" of land plots, which resembled the scales on a fish (*Taxation and Government Finance*, p. 42).

12. *XXZ*, 1:29a. The emperor's approval came in early 1773.

13. Whether *zhuangbao* refers to village headmen or baojia leaders is unclear. See Feuerwerker, *State and Society in Eighteenth-Century China*, p. 98, for the distinction.

14. *XXG*, 3:36a–40a; *XXZ*, 1:17b–18b.

15. The following account is based on *Xiaoshan Lai shih jiapu,* 3:7a–10b; *XXZ,* 1:29a, 4:21a–23a; and *XXG,* 18:10a–b.

16. Compare Smil's account detailing the problems of leeches in reservoirs, wells, and paddies in southern China in the 1980s (*The Bad Earth*, p. 145).

17. This speech is from *XXG,* 18:10a.

18. Yu is referred to as *shenshi* in the sources, although his specific degree is not listed in the gazetteer.

19. The account of Yu's activities is based on *XK.*

20. *XXG,* 3:40b–41b. For an important study of local water control in central China during this period, see Perdue, "Water Control in the Dongting Lake Region," pp. 747–65.

21. *XK,* 2a.

22. *XXG,* 3:40a, 42a.

23. For discussions of the public roles of private elite managers, see Schoppa, *Chinese Elites and Political Change.* See also Rankin, *Elite Activism and Political Transformation in China.*

24. *XXG,* 3:43a.

25. *XXZ,* 2:1a; *XK,* preface, n.p.; and *XXG,* 3:43a.

26. *XK,* 3a.

27. *XK,* preface, n.p. The phrase came, as Benjamin Elman points out, to be used generally as a watchword for unbiased research in the eighteenth century (*From Philosophy to Philology*).

28. A paraphrase, not a translation, of Yu's summary follows.

29. *XK,* 4a–5a.

30. *XK,* 6a–7a. For the Festival of Hungry Ghosts, see Welch, *Practice of Chinese Buddhism*, pp. 109, 125, 185, 188, 200.

31. *XK,* 10a–11b.

32. *XK,* 12a–b.

33. *XK,* 18a–b.

34. *XK,* 28a–b.

35. *XK,* 16a–17a.

36. *XK,* 24b.

37. *XK,* 24a–25b.

38. *XK,* 26a–b.

39. *XK,* 22a–b.

40. The Lian Dike embankment was still used in the 1930s; see *Xiaoshan xiangtu zhi*, p. 45.

41. *Xiaoshan xiangtu zhi*, p. 45.

42. See the biography of Han Zhaopei in *Xiaoshan Yiqiao Han shi jiapu,* 2:21a–b.

43. *XXG,* 33:24b–25a.

44. *XXG,* 7:38a.

45. *XXG,* 5:28b.

46. *XXZ,* 2:1a–b, 4:29a–b.

47. *XXZ,* 4:29b–32b, 6:15a–b.

48. *XXZ,* 4:31a.

49. *XXZ,* 5:21b–23a.

VI. VIEW: *Yewu Mountain, Late 1860s*

1. *XXG*, 1:27b.
2. *XXZ*, 7:40a.
3. See the poem by Wang Mian, which describes it in that way, *XXZ*, 7:34b–35b.
4. *XXZ*, 8:13b–14a.
5. See, e.g., the poems by He Zengyun and Lin Guogui, *XXZ*, 7:34a–b, 38a–40a, for autumn leaves; for love-seeds, see XXZ, 8:14a.
6. See the biography of Xiang Yu from Sima Qian's *Shi Ji* in *Selections from Records of the Historian*, trans. Yang and Yang, pp. 205–37. See also the account from the *History of the Former Han Dynasty* by Ban Gu, ed. Dubs, pp. 62–84.
7. *Selections from the Records of the Historian*, trans. Yang and Yang, p. 220.
8. *XXG*, 32:48a.
9. See *XXZ*, 7:27b, 30a–31a.
10. *XXZ*, 7:34a–b.
11. *XXZ*, 8:7a.
12. *XXZ*, 8:7a.

VI. CHAPTER: *The Beginning of the End: Cataclysm and Silence, 1861–1901*

1. *XXG*, 5:30b.
2. See, e.g., Giquel, *Journal of the Chinese Civil War*, p. 116, and Moule, *New China and Old*, p. 38. An essay by Liu Ts'ui-jung suggests that in two lineages in Xiaoshan many widows did remarry. The great stress on the ethical principle of female chastity, then, comes because it was perceived to be insufficiently honored. See "Demography of Two Chinese Clans," pp. 20–21.
3. *XXG*, 1:32a. Prosper Giquel noted on October 9, 1864, that "yellowish-gold" rice fields in Xiaoshan were ready for harvest (*Journal of the Chinese Civil War*, p. 116).
4. For the Taiping movement as a whole, see Jen, *Taiping Revolutionary Movement*.
5. Schoppa, *Chinese Elites and Political Change*, p. 161.
6. For the military account of the Taipings in Xiaoshan, see *XXG*, 11:9a–10b.
7. See Michael, *Taiping Rebellion*, 3:1241, 1260–61. On Taiping actions in Shaoxing prefecture, see Gu, *Yuezhou jilue*, 6:767–73; Lu Shurong, *Hukou riji*, pp. 789–804; and Yang Derong, *Xiachong ziyu*, pp. 777–83.
8. *XXG*, 11:9a. See also Moule, *New China and Old*, p. 3, and Smil, *Bad Earth*, p. 145.
9. *XXG*, 18:25a and 19:16a–b.
10. *XXG*, 19:13a.
11. See *XXG*, 20. The Taiping allowed their hair to grow rather than shave the front part of their heads and plait the longer hair in the back into a queue.
12. *XXG*, 19:10b.
13. See, e.g., the account of Lai Jue in *Xiaoshan Lai shi jiapu*, 4:42b–43a. Also see the biography of Lai Siyin, *XXG*, 19:13a.
14. The classic study of the formation of militia units is Kuhn, *Rebellion and Its Enemies in Late Imperial China*.

15. *Xiaoshan Shi shi zongpu*, 14: each biography paginated separately.
16. *XXG*, 19:1a–b, 12a.
17. *XXG*, 19:11b–12a.
18. *XXG*, 19:14a.
19. *Xiaoshan Lai shi zongpu*, 3:49a–52b.
20. *XXG*, 11:9b.
21. For an account of the Bao Village effort, see Cole, *The People Versus the Taipings*.
22. See the lists of casualties in *XXG*, juan 20; this list indicates deaths at Bao Village. For the Hua, see *Xiaoshan Yulin Hua shi zongpu*, juan entitled "Biographies." Each biography has separate pagination. Other militia leaders like Wang Miancao and Yang Fengcao who had had some military successes in their home areas also joined the effort at Bao village. See *XXG*, 11:9b–10a, 11a.
23. *XXG*, 11:10a, 19:22a–23a.
24. This account is based on Cole, *The People Versus the Taipings*, pp. 18–21.
25. *XXG*, 19:12a.
26. Rankin, *Elite Activism and Political Transformation*, p. 61; see also Ho, *Studies on the Population of China*, p. 241; Giquel was told by Chinese official Ding Richang that 40 percent of the Zhejiang population was killed (*Journal of the Chinese Civil War*, p. 63). The population records in the Xiaoshan gazetteer are unusable because of their obvious inaccuracy.

Later demographers suggest that the blood-letting temporarily relieved the population pressure on those who remained alive. Population in Shaoxing was further reduced in subsequent decades when large numbers of people emigrated to prefectures north of Hangzhou Bay—an area that experienced even greater destruction than Shaoxing. See Ho, *Studies on the Population of China*, pp. 241*ff*.

27. See, e.g., biographies of Shan Yuting in *Xiaoshan Xihe Shan shi jiapu*, juan 13: each biography paginated separately; of Wu Li'nan in *Xiaoshan Wu shi zongpu*, 2: each biography paginated separately; and of Lin Jisheng in *XXG*, 19:13b.
28. *XXG*, 19:11b.
29. *XXG*, 19:13b.
30. The phrase is Giquel's (*Journal of Chinese Civil War*, p. 65).
31. Moule, *New China and Old*, pp. 179–80.
32. *XXG*, juan 20. The lineages are the Cai, Chen, Han, He, Huang, Jiang, Lai, Lu, Ni, Qu, Tian, Wang, and Zhang.
33. The gazetteer indicates that 2,161 men were killed in the fighting. This figure is certainly too low for all the fighting in the county, but the gazetteer does not explain how it derived these figures.
34. *Xiaoshan Chelijiang Wang shi jiapu*, 3:79a–80b.
35. Kingsmill, "Retrospect of Events on China and Japan," p. 143. Quoted in Wright, *Last Stand of Chinese Conservatism*, p. 122.
36. Duan, *Jinghu zizhuan nianpu*, p. 185.
37. See *XXG*, 7:38a, 33:50b–51a; the biography of Wu Bingyan in *Xiaoshan Wu shi zongpu*, 2: each biography paginated separately; and the report on lineage property in *Xiaoshan Linpu Jiang shi zongpu*, 3:1a–2a.
38. See *XXG*, 7, 8 *passim*.

39. A scourge of vandals, the Taiping set about destroying cultural relics. See the reference to an inscription stone destroyed on the King of Yue Garrison Mountain in *XXZ*, 8:10b. Not all destruction was by rebels: in the fog-enabling victory of Lai Jinfang's militia forces over the Taiping, the Lai lineage militia destroyed bridges so that the Taiping could not cross. Such scorched-earth destruction, while only of brief military value, added substantially to the immense task of reconstruction following the rebellion. See *Xiaoshan Lai shi zongpu*, 3:49a–52b.

40. *XXZ*, 4:32b–34a.

41. This account is based on *XXZ*, 5:21b–23a.

42. The poem is in XXZ, 7:40a–b.

43. *XXG*, 5:30b.

44. *XXG*, 5:30a.

45. *Xiaoshan Chelijiang Wang shi jiapu*, 3:53b–58a.

46. *XXG*, 5:30b.

47. *XXZ*, 8:13a.

48. Duan, *Jinghu zizhuan nianpu*, p. 208.

49. *XXG*, 5:30b.

50. *Xianghu diaocha jihua baogao shu*, pp. 2–3.

51. *XXG*, 3:11b–12a. Haining and Renhe county dikes in Hangzhou prefecture, by contrast, were funded by the imperial court.

52. *Xiaoshan Chelijiang Wang shi jiapu*, 3:53b–58a. See also the biography of He Can, descendant of He Shunbin, in *Xiaoshan Qinyi He shi zongpu*, 3: each biography paginated separately.

53. This discussion is in Duan, *Jinghu zizhuan nianpu*, pp. 209–11, and the biography of Jiang Zhiqi in *Xiaoshan Linpu Jiang shi zongpu*, 3:10b–16a.

54. Shanyin and Guiji had 890,000 mu of irrigated paddy; Xiaoshan had 360,000.

55. It is likely that expectant officials waiting for assignment assumed deputies' responsibilities.

56. This account is from XXZ, 2:2a–3a.

57. *XXZ*, 4:32b–34a.

58. *XXZ*, 2:3b.

59. *XXG*, 5:30b–31a. The phrase is Moule's (*Half a Century in China*, p. 108).

60. *XXZ*, 2:3b.

61. *XXZ*, 2:5a. At the beginning of the lake's history, it had shared a dike with the Qiantang River, hence its use of the river dike fund. See *XXZ*, 4:35b.

62. See Rankin, *Elite Activism and Political Transformation*, chap. 3. I am also indebted to Mark Schwehn for insights and suggestions on the hesitancy of lineage involvement in the late nineteenth century.

63. For Huang Zhongyao's biography, see XXG, 19:22a–23a.

64. For this case, see *XXZ* 2:5a–8a, 4:23a–24b.

65. *XXZ*, 7:43a.

VII. VIEW: *From One View Pavilion on Stone Grotto Mountain, 1925*

1. All these themes and images can be seen in poems by Lai Risheng, Huang Jiugao, Lai Sanping, Cai Zhongguang, Wang Congyan, and Tang Jinzhao. See *XXZ*, 7:7a–9b, 29a–30a.

2. *XXZ*, 7:7b.
3. *XXZ*, 7:9a–b.
4. *XXZ*, 7:29a–b.
5. *XXZ*, 7:29b–30a.
6. For these buildings, see *XXZ*, 5:18a–b, 8:7b, 12a, and *Xiaoshan xiangtu zhi*, p. 56.
7. *XXZ*, 7:41b.
8. *XXZ*, 7:41a.
9. I am indebted to Mark Schwehn for suggestions on the rice and pearl imagery.

VII. CHAPTER: *Fending Off the Scavengers, 1903–1921*

1. See Schoppa, *Chinese Elites and Political Change*. See also Wright, "Introduction: The Rising Tide of Change," pp. 1–63.
2. The proposal is in *XXZ*, 2:8a–11a.
3. *XXZ*, 2:11b–12b.
4. *XXZ*, 2:12b–14a.
5. *XXZ*, 2:34b–37a.
6. *XXZ*, 2:14a.
7. Schoppa, *Chinese Elites and Political Change*, esp. p. 8, chaps. 2, 3, and 5. See also Chang, *Liang Ch'i-ch'ao and Intellectual Transition in China*, pp. 95–100, 154–56.
8. *XXZ*, 2:14a–16a.
9. At times in the proposal Huang assumes a patronizing tone, suggesting that those who opposed his ideas were "rustics" and that even women and children should be able to see the worth of his proposal.
10. These three were Chen Guangqu, Chen Shigui, and Han Dichang.
11. *XXZ*, 2:25b–27a.
12. *XXZ*, 2:27b–29a.
13. *XXZ*, 2:29a–32a.
14. *XXZ*, 2:34b–37a.
15. *XXZ*, 2:37a–b. The estimate of attendance came from the magistrate in his report to the governor.
16. *XXZ*, 2:34b–37a.
17. *XXZ*, 2:34b–37a.
18. *XXZ*, 2:32a–34a.
19. *XXZ*, 2:34b–37a.
20. *XXZ*, 2:37b.
21. *North China Herald*, May 13, 1911, p. 407.
22. *XXG*, 19:23a.
23. See the account in the biography of Wang Qingluo in *Xiaoshan Chelijiang Wang shi jiapu*, 3:85a–87b.
24. *Shi Bao*, December 4, 1911.
25. *XXZ*, 3:1a–2a.
26. *XXZ*, 3:3a–5b.
27. *XXZ*, 3:3a–5b.
28. *XXZ*, 3:6b–11b.
29. *XXZ*, 3:22a–23b.

30. The one exception was Han Xiang of Yiqiao. The other lineages were Hua, Wang, Yu, Li, Kong, and Sun.
31. *XXZ*, 3:13a–b.
32. *XXZ*, 3:13b–14a.
33. *XXZ*, 3:14a–b.
34. *XXZ*, 3:15b–16a.
35. *XXZ*, 3:16b–17b.
36. *XXZ*, 3:23b–26b.
37. The new regulations chose *min* (people) for private ownership, a neutral term, in contrast to *si*, a value-laden term used throughout the lake's history for "private." The latter has the connotation of selfish.
38. *XXZ*, 3:30a–33a.
39. *XXZ*, 3:33a–34b.
40. *XXZ*, 3:35a–36b.
41. *XXZ*, 3:39b–41b.
42. *Zhejiang Shengyihui diyi jie changnian hui yishilu*, vol. 1.
43. *Zhejiang Shengyihui diyi jie changnian hui yishilu*, 3:41b–43b.
44. *Zhejiang Shengyihui diyi jie changnian hui yishilu*, 3:43b–44b.
45. *Zhejiang Shengyihui diyi jie changnian hui yishilu*, 3:35a–36b.
46. *XXZ*, 4:37b–38b.
47. Schoppa, "Politics and Society in Chekiang," p. 78. See also *Shi Bao*, November 30, 1916.
48. *XXZ*, 4:26b.
49. The similarity of this analysis to the picture of Chinese society that emerges from the short stories of Lu Xun, a native of Shaoxing, is noteworthy. The characteristics of Ah Q are immediately brought to mind by the attitudes of people around Xiang Lake. The image of eating the lake and preying on other people suggests Lu's picture of the madman. See "The True Story of Ah Q" and "A Madman's Diary" in Lu Xun, *Selected Works*.

VIII. VIEW: *From One View Pavilion on Genuine Peace Mountain, 1937*

1. *XXZ*, 1:8b–10a.
2. *Xiaoshan xiangtu zhi*, p. 56.
3. *XX* (1985), *Da shiji*, p. 34.
4. This description comes from Wei, "Guanyu Xianghu xiangcun jianshe zhi shangque," p. 10.
5. *Diaocha*, p. 3.

VIII. CHAPTER: *Ends and Means, 1926–1937*

1. *XXX*, 1a.
2. *XXX*, 1b–2b.
3. *XXX*, 4b–5a.
4. *XXX*, 5b–7b, 19a–b.
5. *XXX*, 9a–12b.

6. On the presence of outsiders, see *XXX*, 1b–2b, 12b–15b.
7. *XXX*, 7b–9a, 12b–15b.
8. *XXX*, 15b–17b.
9. *XXX*, 17b–19a.
10. See Schwarcz, *Chinese Enlightenment*, pp. 106–10.
11. *XXX*, 22b–24b.
12. *XXX*, 24b.
13. *XXX*, 24b–28b.
14. *XXX*, 28b–29a.
15. *XXX*, 29a–30b. Xia's main argument rested on the use of *guan* (official) to describe the lake in three disputes out of more than twenty since the Song. These documents clearly used *guan* to stand for *gong* (public).
16. *XXX*, 19b–22b.
17. *XXX*, 32a–35b.
18. *XXX*, 35b–36a.
19. *XXX*, 37a–39a.
20. *XXX*, 39a–40a.
21. *XXX*, 39a–40a.
22. *XX* (1985), 2:93.
23. Schoppa, "Politics and Society in Chekiang," p. 246.
24. *XXX*, 42b–43b.
25. *Shibao*, April 11, 1927, and *Shenbao*, August 8, 1928.
26. *Diaocha*, p. 11.
27. *Diaocha*, preface; see also Yang and Lin, "Xianghu nongcun jianshe shiye zhi gaishu," individual pagination, p. 1.
28. On the establishment of the university, see *Shenbao*, July 20, 1927.
29. *Diaocha*, p. 86.
30. *Diaocha*, pp. 9–10.
31. *Diaocha*, pp. 82, 86–87.
32. *Diaocha*, p. 94.
33. *Diaocha*, pp. 80–81. The agricultural products: rice, wheat, broad beans, tobacco, tea, rape, mulberry and silkworms, the strawberry tree, grapes, turnips, and cabbage; the water products: bream, carp, and three other species of fish, lotus, water chestnut, and water lily; the timber: pine, cypress, fir, catalpa, locust, camphor and *chun*. In conjunction with the project, fish-cultivation would begin in White Horse Lake.
34. *Diaocha*, pp. 92–94.
35. Wei, "Guanyu Xianghu xiangcun jianshe zhi shangque," p. 21.
36. XXJJ, p. 4.
37. See, e.g., *Zhejiang sheng zhengfu gongbao*, May 31, June 24, August 14, October 15, 1927, January 13, February 13, 1928.
38. In a meeting of October 28, 1986, with county officials, the vice-editor of the 1985 Xiaoshan gazetteer argued that the "silence" resulted from the threat of Nationalist military power on the local level.
39. Cao Shu, "Xianghu Dingshancun gaijin xianguang," in *Zhejiang jianshe yuekan*, vol. 10, n. 2, individual pagination, p. 5.

40. Wei, "Guanyu Xianghu xiangcun jianshe zhi shangque," p. 25.
41. XXJJ, p. 4.
42. Cao Shu, "Xianghu Dingshancun gaijin xianguang," p. 5; XXJJ, p. 2.
43. XXJJ, p. 4.
44. *Diaocha*, p. 29.
45. XXJJ, p. 5; Wei, "Guanyu Xianghu xiangcun jianshe zhi shangque," p. 25.
46. This critique is found in XXJJ, pp. 5–11.
47. On cooperatives, see Yang and Lin, "Xianghu nongcun jianshe shiye zhi gaishu," pp. 2–3.
48. *Shenbao*, January 21, 1929.
49. *Shenbao*, June 5, 1930.
50. Gong, "Xianghu nongcun jianshe shengzhong ying you zhi yuanyi shiye," p. 19.
51. Gong, "Xianghu nongcun jianshe shengzhong ying you zhi yuanyi shiye," p. 20.
52. Boorman, ed., *Biographical Dictionary of Republican China*, 3:243–48.
53. As a writer in the school publication *Xiang Lake Life* put it: "Education is the fundamental activity in building the nation. National culture, national spirit both depend on the quality of education." Xuan, "Xiangcun jiaoyude xin lingqu," pp. 7–13.
54. Wang and Chen, "Xianghu shifan chuan jian wushi zhou nian huiyi, " pp. 64, 68–69. See also "Tao Xingzhi yu Xianghu shifan," p. 27, and *XX* (1985), 20:23.
55. Wang and Chen, "Xianghu shifan chuan jian wushi zhou nian huiyi," pp. 68–69.
56. Wang and Chen, "Xianghu shifan chuan jian wushi zhou nian huiyi," p. 66.
57. Wang and Chen, "Xianghu shifan chuan jian wushi zhou nian huiyi," p. 71.
58. Wang and Chen, "Xianghu shifan chuan jian wushi zhou nian huiyi," pp. 69–70.
59. Shen, "Chuangban Xiangbei zhongxin xiaoxue jingguo," pp. 95–97. There were other responses besides fear. East Wang villagers apparently contributed over 700 yuan to build a school without much outside direction and with no disturbances. See Weng and Ye, "Xiangxi zhongxin xiaoxue gaikuang," pp. 92–94.
60. Wang and Chen, "Xianghu shifan chuan jian wushi zhou nian huiyi," p. 71.
61. Wang and Chen, "Xianghu shifan chuan jian wushi zhou nian huiyi," p. 71; *XX* (1985), 20:23.
62. Wang and Chen, "Xianghu shifan chuan jian wushi zhou nian huiyi," p. 75.
63. Wang and Chen, "Xianghu shifan chuan jian wushi zhou nian huiyi," pp. 72–73; *XX* (1985), 20:24.
64. *XX* (1985), *Da shiji*, p. 31.
65. Wang and Chen, "Xianghu shifan chuan jian wushi zhou nian huiyi," p. 74.

IX. VIEW: *From Little Li Mountain, Late 1980s*

1. *XK*, 12a–b.
2. *XX* (1985), 2:97–102.
3. *XX* (1985), 5:11.
4. Interview with Professor You Xiuling, Zhejiang Agricultural College, October 6, 1986.
5. *XX* (1985), 5:11–14.

6. *XX* (1985), 5:25–29.
7. *Xiaoshan xian dimingzhi*, p. 329, and *Xianghu tongxun*, pp. 42–43.
8. *XX* (1985), 13:36–38, 7:61.
9. *XX* (1985), 13:36–38, 7:61.

IX. CHAPTER: *Transmutations, 1937–1986*

1. *XX* (1985,) *Da shiji*, p. 34.
2. *Xiaoshan xian ming sheng jilue* (1934), p. 6.
3. *XX* (1985,) *Da shiji*, pp. 34–35.
4. *XX* (1985), 15:33.
5. *XX* (1985), 20:24.
6. *XX* (1985), *Da shiji*, p. 34.
7. *Xiaoshan gailan*, p. 29.
8. The following account is based on *Xiaoshan gailan*, p. 29; *Xianghu tongxun*, pp. 42–43; and *XX* (1985), *Da shiji*, pp. 36–38.
9. *Xiaoshan gailan*, (1947), p. 29; XX (1985), *Da shiji*, pp. 36, 39.
10. *XX* (1985), 24:48.
11. *XX* (1985), 4:53, 70, 77.
12. *XX* (1985), 4:61.
13. *XX* (1985), 7:1–2.
14. *XX* (1985), 7:29, 32
15. *XX* (1985), 13:11
16. *XX* (1985), *Fulu*, p. 11. For Kong's party leadership, see *Shen Bao*, September 20, 1927.
17. Wei, "Guanyu Xianghu xiangcun jianshe zhi shangque," p. 25.
18. See *Xiaoshan xian canyihui disan ci* (February 1947) and ... *diwu ci* (September 1947).
19. *Dongnan ribao* (Hangzhou), July 29, 1948.
20. This account is based on XX (1985), 16:15. Purchasing rice outside the county was often necessary. See, e.g., *Shen Bao*, July 27, 1927, May 2, 1928.
21. Report of this plan comes from Lou Gengyang, vice-editor of the county gazetteer. The episode emerged in oral interviews Lou conducted for the gazetteer. Written documentation to corroborate this account was not available (interview, October 28, 1986).
22. *XX* (1985), 11:22.
23. *Xiaoshan dizheng*, p. 9.
24. *XX* (1985), *Da shiji*, p. 43.
25. Jin, "Zhedong kangri genjudi de chuangjian," p. 117.
26. *XX* (1985), 20:25, *Da shiji*, pp. 39–43.
27. *XX* (1985), *Da shiji*, pp. 43–47.
28. *XX* (1985), *Da shiji*, p. 45.
29. *XX* (1985), 16:3.
30. *XX* (1985), *Da shiji*, pp. 46–47.
31. *XX* (1985), 24:56–65 *passim*.
32. *Xiaoshan xiangtu zhi*, pp. 92–93.

33. *XX* (1985), 4:27.
34. Discussions held in the Xiaoshan county offices, October 28, 1986,
35. Corroborated in interview with You Xiuling, October 6, 1986.
36. *XX* (1985), *Fulu*, pp. 11–14, and *Xiaoshan xian dimingzhi*, p. 514.
37. The One View was still intact in 1947 (*Xiaoshan gailan*, p. 21; cf. *XX* [1985], 21:27).
38. *Xiaoshan xian dimingzhi*, p. 528.
39. *Xiaoshan xian dimingzhi*, p. 496.
40. Interview with Chen Qiaoyi, September 22, 1986.
41. *XX* (1985), 24:47–48.
42. *XX* (1985), *Da shiji*, p. 59.
43. *XX* (1985), *Da shiji*, p. 59.
44. *XX* (1985), *Da shiji*, p. 60.
45. *XX* (1985), *Da shiji*, pp. 58, 61.
46. See Meisner, *Mao's China*, pp. 244–47.
47. *XX* (1985), *Da shiji*, p. 61.
48. See, e.g., *XX* (1985), 7:40–65, and interview with Chen Qiaoyi, September 1986.
49. *XX* (1985), 1:30.
50. *XX* (1985), *Da shiji*, pp. 70–72.
51. *XX* (1985), 15:22.
52. *XX* (1985), *Da shiji*, p. 84.
53. *XX* (1985), p. 89.
54. *XX* (1985), 14:20–23, 15:18–21.
55. This account is based on *XX* (1985), 7:22, 8:25, and *Fulu*, pp. 12–14.
56. Meeting with county officials, Xiaoshan county, October 28, 1986.
57. Map of Xiaoshan, 1986; interview with Que Weimin, September 18, 1986.
58. *XX* (1985), 21:37.
59. *XX* (1985), 4:69.

Bibliography

Ayscough, Florence. "The Chinese Idea of a Garden." *Chinese Journal of Science and Arts* 1:1 (January 1923): 15–22.

Ban Gu. *History of the Former Han Dynasty*. Trans. and ed. Homer Dubs. Baltimore: Waverly Press, 1938.

Boorman, Howard L., ed. *Biographical Dictionary of Republican China*. 4 vols. New York: Columbia University Press, 1970.

Cao Shu. "Xianghu Dingshancun gaijin xiankuang" (The current progress at Xiang Lake's Ding Village), *Zhejiang jianshe yuekan* 10:2 (1936): 5.

Chang Hao. *Liang Ch'i-ch'ao and Intellectual Transition in China, 1890–1907*. Cambridge: Harvard University Press, 1971.

Chen, Kenneth K. S. *Buddhism in China*. Princeton: Princeton University Press, 1964.

Chen Qiaoyi. "Lun lishi shiqi Ning-Shao pingyuandi hupo yanbian" (On changes in the lakes of the Ning-Shao Plain throughout History), *Dili yanjiu* 3:3 (September 1984): 29–43.

———. "Lun lishi shiqi Puyangjiang xiayude hedao bianqian" (On changes in the channel of the Lower Puyang River throughout history), *Lishi dili* 1 (1981): 65–79.

Chinese Economic Bulletin, 1927–1937.

Cole, James H. *The People Versus the Taipings: Bao Lisheng's "Righteous Army of Dongan."* Berkeley: Center for Chinese Studies, 1981.

———. *Shaohsing*. Tucson: University of Arizona Press, 1986.

Daoguang Guiji xianzhi (Gazetteer of Guiji county in the Daoguang period), 1821–1851 (specific date uncertain).

De Groot, J. J. M. *The Religious System of China*. 3 vols. Reprint, Taibei: Cheng Wen, 1972.

Dennerline, Jerry. *The Chia-ting Loyalists.* New Haven and London: Yale University Press, 1981.

———. "Marriage, Adoption, and Charity in the Development of Lineages in Wu-hsi from Sung to Ch'ing." In Patricia Buckley Ebrey and James L. Watson, eds., *Kinship Organization in Late Imperial China, 1000–1940.* Berkeley: University of California Press, 1986.

Dongnan ribao (Southeast Daily) (Hangzhou), 1948.

Duan Guangqing. *Jinghu zizhuan nianpu* (Autobiography of Duan). Reprint, Taibei: n.d.

Eberhard, Wolfram. *The Local Cultures of South and East China.* Leiden: E. J. Brill, 1968.

Ebrey, Patricia Buckley. "The Early Stages in the Development of Descent Group Organization." In Patricia Buckley Ebrey and James L. Watson, eds., *Kinship Organization in Late Imperial China, 1000–1940.* Berkeley: University of California Press, 1986.

Elman, Benjamin. *From Philosophy to Philology.* Cambridge: Harvard University Press, 1984.

Elvin, Mark. "Market Towns and Water Ways: The County of Shanghai from 1480 to 1910." In G. William Skinner, ed., *The City in Late Imperial China.* Stanford: Stanford University Press, 1977.

———. *The Pattern of the Chinese Past.* Stanford: Stanford University Press, 1973.

Feuerwerker, Albert. *State and Society in Eighteenth-Century China: The Ch'ing Empire in Its Glory.* Ann Arbor: Center for Chinese Studies, 1976.

Fitzpatrick, Merrilyn. "Local Administration in Northern Chekiang and the Response to the Pirate Invasions of 1553–1556." Ph.D. diss., Australian National University, 1976.

Fogel, Joshua A. *Politics and Sinology: The Case of Naitō Konan (1866–1934).* Cambridge: Harvard University Press, 1984.

Franke, Herbert. "Siege and Defense of Towns in Medieval China." In Frank A. Kierman, Jr., and John K. Fairbank, eds., *Chinese Ways in Warfare.* Cambridge: Harvard University Press, 1974.

Freedman, Maurice. *Family and Kinship in Chinese Society.* Stanford: Stanford University Press, 1970.

Gernet, Jacques. *Daily Life in China on the Eve of the Mongol Invasions, 1250–1276.* New York: MacMillan, 1962.

Giquel, Prosper. *A Journal of the Chinese Civil War, 1864.* Ed. Steven A. Leibo. Honolulu: University of Hawaii Press, 1985.

Gong Ying. "Xianghu nongcun jianshe shengzhong ying you zhi yuanyi shiye" (The necessary undertaking of the arts of gardening in Xiang Lake's rural reconstruction), *Zhejiang nongye tuiguang* 11:12 (1936): 18–21.

Goodrich, L. Carrington, ed. *Dictionary of Ming Biography*. 2 vols. New York: Columbia University Press, 1976.

Grimm, Tilemann. "Ming Educational Intendants." In Charles Hucker, ed., *Chinese Government in Ming Times*. New York: Columbia University Press, 1969.

Gu Yue Yinmingshi (pseud.). "Yuezhou jilue" (An account of Shaoxing prefecture) In Xiang Da, ed., *Taiping Tianguo*, vol. 6. Shanghai, 1952.

Han Yanmen. "Xianghu xiangcun jianshe jihua dagong cao'an" (A draft of the planning principles of Xiang Lake's reconstruction), *Zhejiang nongye tuiguang* 11:12 (August 1936) : 4–11.

Hartwell, Robert. "Demographic, Political, and Social Transformation of China, 750–1550." *Harvard Journal of Asiatic Studies* 42:2 (December 1982): 365–442.

Hayes, James. "Specialists and Written Materials in the Village World." In David Johnson, Andrew J. Nathan, and Evelyn S. Rawski, eds., *Popular Culture in Late Imperial China*. Berkeley: University of California Press, 1985.

Hazelton, Keith. "Patrilines and the Development of Localized Lineages: The Wu of Hsiu-ning City, Hui-chou, to 1528." In Patricia Buckley Ebrey and James L. Watson, eds., *Kinship Organization in Late Imperial China, 1000–1940*. Berkeley: University of California Press, 1986.

Higgins, Roland. "Pirates in Gowns and Caps: Gentry Law-Breaking in the Mid-Ming," *Ming Studies* 10 (Spring 1980): 30–36.

Hirth, Friedrich. *The Ancient History of China*. New York: Columbia University Press, 1911.

Ho Ping-ti. *The Ladder of Success in Imperial China*. New York: Columbia University Press, 1962.

———. *Studies on the Population of China, 1368–1953*. Cambridge: Harvard University Press, 1959.

Hsieh, Andrew C. K., and Jonathan D. Spence. "Suicide and the Family in Pre-Modern Chinese Society." In Arthur Kleinman and Tsung-yi Lin, eds., *Normal and Abnormal Behavior in Chinese Culture*. Dordrecht: D. Reidel, 1981.

Huang, Ray. *1587, A Year of No Significance: The Ming Dynasty in Decline*. New Haven and London: Yale University Press, 1981.

———. "Fiscal Administration during the Ming Dynasty." In Charles Hucker, ed., *Chinese Government in Ming Times*. New York: Columbia University Press, 1969.

———. *Taxation and Government Finance in Sixteenth-Century Ming China*. Cambridge: Cambridge University Press, 1974.

Hummel, Arthur, ed. *Eminent Chinese of the Ch'ing Period*. 2 vols. Washington, D.C.: U.S. Government Printing Office, 1943.

Hwang Kwang-kuo. "Face and Favor: The Chinese Power Game." Manuscript.

Hymes, Robert P. "Marriage, Descent Groups, and the Localist Strategy in Sung and Yuan Fu-chou." In Patricia Buckley Ebrey and James L. Watson, eds., *Kinship Organization in Late Imperial China, 1000–1940*. Berkeley: University of California Press, 1986.

———. *Statesmen and Gentlemen: The Elite of Fu-chou, Chiang-hsi, in Northern and Southern Sung*. Cambridge: Cambridge University Press, 1986.

Jen Yu-wen. *The Taiping Revolutionary Movement*. New Haven: Yale University Press, 1973.

Jin Pusen. "Zhedong KangRi genjudi de chuangjian" (The establishment of the anti-Japanese base area in eastern Zhejiang), *Hangzhou daxue xuebao* 15:3 (September 1985): 116–23.

Kao Yu-kung. "The Fang La Rebellion." *Harvard Journal of Asiatic Studies* 24 (1962–1963): 27–61.

Kenneson, James. "China Stinks." *Harper's* (April 1982): 13–19.

Kingsmill, Thomas W. "Retrospect of Events in China and Japan during the Year 1865." *Journal of the North China Branch of the Royal Asiatic Society* 2 (1865): 143.

Kuhn, Philip. *Rebellion and Its Enemies in Late Imperial China*. Cambridge: Harvard University Press, 1970.

Liu Ts'ui-jung. "The Demography of Two Chinese Clans in Hsiao-shan, Chekiang, 1650–1850. In Susan B. Hanley and Arthur P. Wolf, eds., *Family and Population in East Asian History*. Stanford: Stanford University Press, 1985.

Lu Shurong. "Hukou riji" (Diary from the tiger's month). In Xiang Da, ed., *Taiping Tianguo*. Vol. 6. Shanghai, 1952.

Lu Xun. *Selected Works*. Trans. Yang Xianyi and Gladys Yang. Beijing: Foreign Languages Press, 1956.

McKnight, Brian E. *Village and Bureaucracy in Southern Sung China*. Chicago: University of Chicago Press, 1971.

Mao Qiling. *He yushi xiaozi ci zhufuwei lu* (A record of the re-establishment of the shrine to Censor He and his filial son). N. p.: n.d.

———. *Xianghu shuili zhi* (Gazetteer of Xiang Lake's conservancy). N.p.: n.d.

Maspero, Henri. *China in Antiquity*. Trans. Frank A. Kierman, Jr. Amherst: University of Massachusetts Press, 1978.

Meisner, Maurice. *Mao's China*. New York: Free Press, 1977.

Meskill, John. "Academies and Politics in the Ming Dynasty." In Charles Hucker, ed., *Chinese Government in Ming Times*. New York: Columbia University Press, 1969.

Michael, Franz. *The Taiping Rebellion.* 3 vols. Seattle: University of Washington Press, 1971.

Mihelic, Mira. "Polders and the Politics of Land Reclamation in Southeast China during the Northern Sung Dynasty." Ph.D. diss., Cornell University, 1979.

Moore, Commander. "The Bore of the Tsien-tang Kiang (Hang-chau Bay)," *Journal of the China Branch of the Royal Asiatic Society* 23 (1889): 180–247.

Morita, Akira. "Shinmatsu Shōko ni tsuite no ichi mondai" (The case of Xiang Lake at the end of the Qing dynasty), *Chūgoku kanken ronsetsu shiryō* 8:4 (1967): 39–63.

———. *Shindai suirishi kenkyū* (A study of the history of water conservancy in the Qing period). Tokyo, 1974.

Morris, Edwin T. *The Gardens of China: History, Art, and Meanings.* New York: Charles Scribner's Sons, 1983.

Moule, Arthur Evans. *Half a Century in China.* London: Hodder and Stoughton, 1911.

———. *New China and Old.* London: Seeley, 1891.

Nickum, James E. *Water Management Organization in the People's Republic of China.* Armonk, N.Y.: M. E. Sharpe, 1981.

North China Herald, 1905–1916.

Olson, Mancur. *The Logic of Collective Action.* Cambridge: Harvard University Press, 1965.

Perdue, Peter. "Water Control in the Dongting Lake Region during the Ming and Qing Periods." *Journal of Asian Studies* 41:4 (August 1982): 747–65.

Rankin, Mary Backus. *Elite Activism and Political Transformation in China: Zhejiang Province, 1865–1911.* Stanford: Stanford University Press, 1986.

Rawski, Evelyn S. "Economic and Social Foundations of Late Imperial Culture." In David Johnson, Andrew J. Nathan, and Evelyn Rawski, eds., *Popular Culture in Late Imperial China.* Berkeley: University of California Press, 1985.

Schoppa, R. Keith. *Chinese Elites and Political Change.* Cambridge: Harvard University Press, 1982.

———. "Politics and Society in Chekiang, 1907–1927: Elite Power, Social Control, and the Making of a Province." Ph.D. diss., University of Michigan, 1975.

Schwarcz, Vera. *The Chinese Enlightenment.* Berkeley: University of California Press, 1986.

Selections from Records of the Historian. Trans. Yang Hsien-yi and Gladys Yang. Beijing: Beijing Foreign Languages Press, 1979.

Shaoxing fuzhi (Gaxetteer of Shaoxing prefecture), 1586.

Shaoxing xianzhi ziliao diyi ji (Compilation of materials for a gazetteer of Shaoxing county), 1937.

Shen Bao, 1923–1930.

Shen Lin. "Chuangban Xiangbei zhongxin xiaoxue jingguo" (The founding of the North Xiang Lake Central Elementary School), *Xianghu shenghuo* 6 (June 1929): 95–97.

Shi Bao, 1909–1927.

Shiba Yoshinobu. *Commerce and Society in Sung China*, Trans. Mark Elvin. Ann Arbor: Center for Chinese Studies, 1970.

———. "*Xianghu shuili zhi* to *Xianghu kaolue*—Sekkō Shōsanken Shōko no suiri shimatsu" (*Gazetteer of Xiang Lake's conservancy* and *A Summary of the Investigation into Xiang Lake*—The circumstances of the conservancy of Xiang Lake). In *Satō hakase taikan kinen: Chūgoku suirishi ronsō* (Collection commemorating the retirement of Professor Satō: Essays on the history of water conservancy in China). Tokyo: Kokusho Kankōkai, 1985.

Smil, Vaclav. *The Bad Earth*. Armonk, N.Y.: M. E. Sharpe, 1984.

So Kwan-wai. *Japanese Piracy in Ming China during the Sixteenth Century*. East Lansing: Michigan State University Press, 1975.

Struve, Lynn. *The Southern Ming, 1644–1662*. New Haven and London: Yale University Press, 1984.

Stuermer, John R. "Polder Construction and the Pattern of Land Ownership in the Tai-hu Basin during the Southern Sung Dynasty." Ph.D. diss. University of Pennsylvania, 1980.

"Tao Xingzhi yu Xianghu shifan" (Tao Xingzhi and the Xiang Lake Normal School), *Zhejiang wenshi ziliao* 3 (1981): 26–28.

Wakeman, Frederick, Jr. *The Great Enterprise: The Manchu Reconstruction of Imperial Order in Seventeenth-Century China*. 2 vols. Berkeley: University of California Press, 1985.

Waley, Arthur, trans. *The Analects of Confucius*. London: George Allen Unwin, 1938.

Wang Canyuan and Chen Dingshun. "Xianghu shifan chuanjian wushi zhou nian huiyi" (Remembrances of the founding of the Xiang Lake Normal School fifty years ago), *Zhejiang wenshi ziliao* 13 (September 1978): 64–75.

Watson, Burton. *Ssu-ma Ch'ien, Grand Historian of China*. New York: Columbia University Press, 1958.

Watson, James L. "Anthropological Overview: The Development of Chinese Descent Groups." In Patricia Buckley Ebrey and James L. Watson, eds., *Kinship Organization in Late Imperial China, 1000–1940*. Berkeley: University of California Press, 1986.

Watt, John R. "The Yamen and Urban Administration." In G. William

Skinner, ed., *The City in Late Imperial China*. Stanford: Stanford University Press, 1977.

Wei Songtang. "Guanyu Xianghu xiangcun jianshe zhi shangque" (Concerning consultations about the rural reconstruction of Xiang Lake), *Zhejiang jingji* (1947): 8–12, 19–25.

Welch, Holmes. *The Practice of Chinese Buddhism, 1900–1950*. Cambridge: Harvard University Press, 1967.

Weng Yanzhen and Ye Lunshu. "Xiangxi zhongxin xiaoxue gaikuang" (West Xiang Lake Central Elementary School), *Xianghu shenghuo* 6 (June 1929): 92–94.

Wiens, Mi Chu. "Changes in the Fiscal and Rural Control Systems in the Fourteenth and Fifteenth Centuries." *Ming Studies* 3 (Fall 1975): 53–67.

Will, Pierre-Etienne. "The Occurrences of, and Responses to, Catastrophes and Economic Change in the Lower and Middle Yangtze, 1500–1850." Paper presented at the Conference on Spatial and Temporal Trends and Cycles in the Chinese Economy, 980–1980, Bellagio, Italy, August 1984.

Worster, Donald. *Nature's Economy*. Cambridge: Cambridge University Press, 1977.

Wright, Arthur F. "The Cosmology of the Chinese City." In G. William Skinner, ed., *The City in Late Imperial China*. Stanford: Stanford University Press, 1977.

Wright, Mary. "Introduction: The Rising Tide of Change." In Mary Wright, ed., *China in Revolution*. New Haven: Yale University Press, 1968.

———. *The Last Stand of Chinese Conservatism*. Stanford: Stanford University Press, 1957.

Xianghu diaocha jihua baogao shu (Report on the investigation of Xiang Lake), 1927.

Xianghu tongxun (Xiang Lake report) 13 (July 7, 1940).

Xiaoshan Chelijiang Wang shi jiapu (Genealogy of the Wang lineage at Chelijiang, Xiaoshan county), 1917.

Xiaoshan Daishang Huang shi jiapu (Genealogy of the Huang lineage of Daishang, Xiaoshan County), 1895.

Xiaoshan dizheng (Xiaoshan land administration) (January 1, 1949): 9.

Xiaoshan gailan (Xiaoshan overview). Hangzhou, 1947.

Xiaoshan Lai shi jiapu (Genealogy of the Lai lineage of Xiaoshan county), 1922.

Xiaoshan Lai shi zongpu (Genealogy of the Lai lineage of Xiaoshan county), 1900.

Xiaoshan Linpu Jiang shi zongpu (Genealogy of the Jiang lineage at Linpu, Xiaoshan county), 1908.

Xiaoshan Qinyi He shi zongpu (Genealogy of the He lineage at Qinyi, Xiaoshan county), 1893.

Xiaoshan Shi shi zongpu (Genealogy of the Shi lineage of Xiaoshan county), 1892.

Xiaoshan Wu shi zongpu (Genealogy of the Wu lineage of Xiaoshan county), 1904.

Xiaoshan xian canyihui, disan ci (Records of the Xiaoshan county board, the third session) (February 1947).

Xiaoshan xian canyihui, diwu ci (Records of the Xiaoshan county board, the fifth session) (September 1947).

Xiaoshan xian dimingzhi (Gazetteer of place names in Xiaoshan county). Xiaoshan, 1984.

Xiaoshan xian ming sheng jilue (A summary of famous scenic sites in Xiaoshan county), 1934.

Xiaoshan xiangtu zhi (A gazetteer of the Xiaoshan locality). Hangzhou, 1931.

Xiaoshan xianzhi (Gazetteer of Xiaoshan county), 1693.

Xiaoshan xianzhi (Gazetteer of Xiaoshan county), 1751.

Xiaoshan xianzhi (Gazetteer of Xiaoshan county), 1985.

Xiaoshan xianzhi gao (Draft of a gazetteer of Xiaoshan county), 1935.

Xiaoshan Xihe Shan shi jiapu (Genealogy of the Shan lineage at Xihe, Xiaoshan county), 1932.

Xiaoshan Yiqiao Han shi jiapu (Genealogy of the Han lineage at Yiqiao, Xiaoshan county), 1932.

Xiaoshan Yulin Hua shi zongpu (Genealogy of the Hua lineage at Yulin, Xiaoshan county), 1916.

Xiaoshan Zhao jiapu (Genealogy of the Zhao lineage of Xiaoshan county), 1896.

Xihe Xiansheng zhuan (Biography of Mr. West River [Mao Qiling]). N.p.: n.d.

Xu Fanglie. *Zhedong jilue* (Record of events in eastern Zhejiang). Reprint, Taibei, 1968.

Xu Mianzhi. *Bao Yue lu* (Record of the protection of Yue). N.p.: n.d.

Xuan Shiqian. "Xiangcun jiaoyude xin lingqu" (The new realm of rural education), *Xianghu shenghuo* 12 (December 1929): 7–13.

Yang, C. K. *Religion in Chinese Society*. Berkeley: University of California Press, 1961.

Yang Cengsheng and Lin Yuejing. "Xianghu nongcun jianshe shiye zhi gaishu" (A summary of the effort at rural reconstruction at Xiang Lake), *Zhejiang jianshe yuekan* 10:2 (1936): 22–25.

Yang Derong. "Xiachong ziyu" (Insects of summer speak for themselves). In Xiang Da, ed., *Taiping Tianguo*. Vol. 6. Shanghai, 1952.

Yule, Sir Henry, trans. *The Book of Ser Marco Polo*. Ed. Henri Cordier. London: John Murray, 1921.
Yu Shida. *Xianghu kaolue* (A summary of the investigation into Xiang Lake). 1798.
Zelin, Madeleine. *The Magistrate's Tael.* Berkeley: University of California Press, 1984.
Zhejiang sheng zhengfu gongbao (Zhejiang provincial government gazette), 1927–1930.
Zhejiang shengyihui diyi jie changnian hui yishilu (A record of the deliberations of the first session of the annual meeting of the Zhejiang Provincial Assembly). 1916.
Zhejiang shuiliju niankan (The Annual of the Zhejiang Provincial Conservancy Board), 1929.
Zhou Yizao, ed. *Xiaoshan Xianghu Xuzhi* (A continuation of the gazetteer of Xiang Lake), 1926.
———. *Xiaoshan Xianghu zhi* (A history of Xiaoshan's Xiang Lake), 1925.

Index

Printed in the United States
38334LVS00004B/156